Mineral Deposits and the Evolution of the Biosphere

H. D. Holland and M. Schidlowski
Editors

Report of the Dahlem Workshop on
Biospheric Evolution and Precambrian Metallogeny
Berlin 1980, September 1-5

Rapporteurs:
S. M. Awramik · A. Button · J. H. Oehler · N. Williams

Program Advisory Committee:
S. M. Awramik · A. Babloyantz · P. Cloud · G. Eglinton
H. L. James · C. E. Junge · I. R. Kaplan · S. L. Miller
M. Schidlowski · P. H. Trudinger

Springer-Verlag Berlin Heidelberg New York 1982

Copy Editor: M. Cervantes-Waldmann
Photographs: E. P. Thonke

With 4 photographs, 41 figures, and 9 tables

ISBN-13:978-3-642-68465-4 e-ISBN-13:978-3-642-68463-0
DOI: 10.1007/978-3-642-68463-0

CIP-Kurztitelaufnahme der Deutschen Bibliotheken:

Mineral deposits and the evolution of the biosphere: report of the Dahlem Workshop on Biospheric Evolution and Precambrian Metallogeny, Berlin 1980, September 1-5 / H. D. Holland and M. Schidlowski, eds. Rapporteurs: S. M. Awramik ... [Dahlem Konferenzen, Held and publ. on behalf of the Stifterverb. für d. Dt. Wiss. Sponsored by Dt. Forschungsgemeinschaft ...]. – Berlin; Heidelberg; New York: Springer, 1982.

(Physical and chemical sciences research reports; 3)

(Dahlem workshop reports)

NE: Holland, Heinrich D. [Hrsg.]; Awramik, Stanley M. [Mitverf.]; Workshop on Biospheric Evolution and Precambrian Metallogeny <1980, Berlin, West>; Dahlem Konferenzen; GT

Softcover reprint of the hardcover 1st edition 1982

2131/3014 – 5 4 3 2 1 0

Table of Contents

BIOLOGICAL PROCESSES AND THE FORMATION OF MINERAL DEPOSITS

The Dahlem Konferenzen

DIRECTOR:
Silke Bernhard, M.D.

FOUNDATION:
Dahlem Konferenzen was founded in 1974 and is supported by the Stifterverband für die Deutsche Wissenschaft, the Science Foundation of German Industry, in cooperation with the Deutsche Forschungsgemeinschaft, the German Organization for Promoting Fundamental Research, and the Senate of the City of Berlin.

OBJECTIVES:
The task of Dahlem Konferenzen is:

to promote the interdisciplinary exchange of scientific information and ideas,

to stimulate international cooperation in research, and

to develop and test different models conducive to more effective scientific meetings.

AIM:
Each Dahlem Workshop is designed to provide a survey of the present state of the art of the topic at hand as seen by the various disciplines concerned, to review new concepts and techniques, and to recommend directions for future research.

PROCEDURE:
Dahlem Konferenzen approaches internationally recognized scientists to suggest topics fulfilling these criteria and to propose members for a Program Advisory Committee, which is responsible for the workshop's scientific program. Once a year, the topic suggestions are submitted to a scientific board for approval.

TOPICS:
The workshop topics should be:

of contemporary international interest,

timely,

interdisciplinary in nature, and

problem-oriented.

PARTICIPANTS:
The number of participants is limited to 48 for each workshop, and they are selected exclusively by a Program Advisory Committee. Selection is based on international scientific reputation alone and is independent of national considerations, although a balance between Europeans and Americans is desirable. Exception is made for younger German scientists for whom 10% of the places are reserved.

THE DAHLEM WORKSHOP MODEL:
A special workshop model has been developed by Dahlem Konferenzen, the *Dahlem Workshop Model*. The main work of the workshop is done in four small, interdisciplinary discussion groups, each with 12 members. Lectures are not given.

Some participants are asked to write background papers providing a review of the field rather than a report on individual work. These are circulated to all participants 4 weeks before the meeting with the request that the papers be read and questions on them formulated *before* the workshop, thus providing the basis for discussions.

During the workshop, each group prepares a Report reflecting the essential points of its discussions, including suggestions for future research needs. These reports are distributed to all participants at the end of the workshop and are discussed in plenum.

PUBLICATION:
The Dahlem Workshop Reports contain:

the Chairperson's introduction,

the background papers, and

the Group Reports.

Mineral Deposits and the Evolution of the Biosphere, eds. H.D. Holland and M. Schidlowski, pp. 1-4. Dahlem Konferenzen, 1982.
Berlin, Heidelberg, New York: Springer-Verlag.

Introduction

H. D. Holland* and M. Schidlowski**
*Dept. of Geological Sciences, Harvard University
Cambridge, MA 02138, USA
**Max-Planck-Institut für Chemie, 6500 Mainz, F. R. Germany

The Dahlem Conference on Mineral Deposits and the Evolution of the Biosphere was born in 1978 at the Ram Hotel in Jerusalem. One of us (M.S.) had the good fortune to be breakfasting there with Dr. Silke Bernhard during the course of the 10th International Sedimentology Congress. The conversation turned to Project 157 of the International Geological Correlation Program, an interdisciplinary project which is investigating the relationships between early organic evolution, mineral deposits, and energy resources. An international conference for specialists in these areas appealed to both of us, and the 23rd Dahlem Conference grew out of our common interest. The topic of the conference gradually evolved into something that was both broader and more restrictive than the charter of the IGCP Project 157. It was broader, because it included not only the early phases but the entire time span of organic evolution; it was more restrictive, because energy resources were not considered in any serious way. As it turned out, both changes were helpful, and the conference as a whole was a great success.

In some sense this success was assured by the choice of topic. Sedimentary ores are probably the most important and most diverse group of mineral deposits. Processes at the interface of the land

and the atmosphere, within river systems, and within the oceans account for nearly all of the ores of aluminum, much of the nickel, a good deal of the gold, uranium, and gem minerals, most of the phosphate, limestone, gypsum, anhydrite, halite, and potash, most of the major iron ore deposits, and for large potential reserves of manganese, copper, nickel, cobalt, molybdenum, and vanadium. Biological processes are directly involved in the formation of some of these sedimentary ores, indirectly in the formation of others, and weakly or not at all in the formation of the remainder. In some instances, the involvment is well established; phosphorites, limestones, and black shales are examples of deposits in this group. In others the involvement of biological factors is still debated; banded iron formations, manganese nodules, and Precambrian detrital uranium ore deposits are members of this group. Not unexpectedly, the course of the Conference was influenced by these present uncertainties and by the uncertainties that cloud our understanding of the origin of life on Earth, the evolution of the biosphere, and its influence on the history of the atmosphere and oceans.

The diversity of opinion and expertise among members of the Dahlem Conference surfaced most dramatically in the debate concerning the existence of fossil remains in the ancient metasedimentary rocks at Isua in West Greenland. In a more subtle manner it has shaped the revisions of the background papers since the conclusion of the conference. These papers have been divided into three groups. The first group is concerned with past and present microbial processes and ecosystems; the second group deals with the morphological and chemical record of the Precambrian biosphere; the third group explores the relationships between biological processes and the formation of mineral deposits. These papers and the state of the art reports that follow them bring together data and ideas from fields that have traditionally not been considered to be parts of a single whole. It is obvious that this fusion has potential implications for biology, paleontology, oceanography, and mineral exploration. The guidelines for mineral exploration that have developed to date are still rather obscure, but they will probably be clarified by the continued cooperative efforts of scientists such as those in the group that met in Berlin last September.

The staff of Dahlem Konferenzen is responsible for making the meeting of this group memorably pleasant and pleasantly memorable. Dr. Bernhard's gifts of charm, organizational skill, and administrative toughness assured that the conference was run elegantly, smoothly, and decisively, even down to the choice of editors for this volume. Marie Cervantes-Waldmann performed minor miracles extracting manuscripts gently but persistently from the authors and in turning the typescripts into a book. The other staff members of Dahlem Konferenzen were unfailingly helpful even under trying circumstances. They will be well remembered by all who were fortunate enough to be asked to Berlin for the first week in September, 1980.

Mineral Deposits and the Evolution of the Biosphere, eds. H.D. Holland and M. Schidlowski, pp. 5-30. Dahlem Konferenzen, 1982.
Berlin, Heidelberg, New York: Springer-Verlag.

Microbial Processes in the Sulfur Cycle Through Time

H. G. Trüper
Institut f. Microbiologie, Rheinische Friedrich-Wilhelms-Universität, 5300 Bonn 1, F. R. Germany

Abstract. Two microbial processes are involved in the sulfur cycle of the earth's biosphere: anoxic dissimilatory sulfur oxidation by phototrophic bacteria and dissimilatory sulfate reduction by sulfate-reducing bacteria. In the presence of oxygen at chemoclines and redoxclines dissimilatory sulfur oxidation by chemolithotrophic bacteria (Thiobacillus, Beggiatoa, and others) occurs. In addition, dissimilatory sulfur reducing bacteria participate in the sulfur cycle. The processes of sulfur assimilation and of microbial liberation of sulfur compounds by decomposition of organic materials are of secondary geomicrobiological importance. Although functioning in opposite directions, the enzymatic equipment of the bacteria involved in dissimilatory sulfur transformations shows a surprising degree of uniformity. Possible phylogenetic relationships between the bacteria involved are discussed on the basis of recent findings in the field of chemotaxomony.

INTRODUCTION

A microbiologist's view on "microbial processes in the sulfur cycle through time" is determined by his knowledge of contemporary microorganisms. This article is therefore mainly a summary of present knowledge of processes in various kinds of pertinent environments rather than a "history of the sulfur cycle" per se. Modern chemotaxonomy has, however, started to supply us with phylogenetic information from which we may at least glean some insight into the evolution of bacteria involved in the biological processes of the sulfur cycle.

In 1925 Baas-Becking (1) coined the term "sulfuretum" for "the natural ecological community" of sulfur bacteria which "is a miniature cycle in itself." A sulfuretum is therefore the ecological community of sulfide-oxidizing and sulfate-reducing bacteria which together catalyze a biological sulfur cycle. Biological ecosystems connected with sulfureta have been termed "die sapropelische Lebewelt" (29) and "the sulfide biome" (12). Baas-Becking's definition of the "sulfuretum" does not imply that such an ecosystem must be free of molecular oxygen; however, a fully oxygenated environment cannot maintain a sulfuretum, because dissimilatory sulfate reduction is a strictly anoxic process. A net input of reduced sulfur compounds from sulfur springs or of sulfate from seawater or salt springs may occur.

Sulfureta are either confined to sediments, as is generally the case at the ocean bottom, but they may also include more or less of the water column; the Black Sea and meromictic lakes are examples of such sulfureta. Sulfureta such as these are in a rather steady-state; others, such as tidal flats, salts marshes, and holomictic lakes, are regularly disturbed by tidal or annual changes. Sulfureta even exist in environments of extreme stress, such as the extremely saline and alkaline Egyptian Wadi Natrun (18,19).

PROCESSES AND MICROORGANISMS OF THE BIOLOGICAL SULFUR CYCLE

Anoxic Microbial Oxidation of Sulfur Compounds

Dissimilatory anoxic oxidation of reduced sulfur compounds depends upon the presence of light or nitrate and of a suitable carbon source (at least CO_2). It is carried out by photolithotrophic bacteria (Chlorobiaceae, Chloroflexaceae, Chromatiaceae, Rhodospirillaceae, or cyanobacteria) or by anaerobic chemolithotrophic nitrate respirers (Thiobacillus denitrificans). All of these bacteria derive electrons from sulfide (or elemental sulfur or thiosulfate) that are needed for nicotinamide adenine dinucleotide (NAD) reduction either via photosynthetic noncyclic electron flow or via adenosine triphosphate (ATP)-dependent reverse electron flow (52). The Chlorobiaceae (green sulfur bacteria) are obligate photolithoautotrophs and strict anaerobes. They lack the reductive pentose phosphate cycle (Calvin cycle), and CO_2 is fixed via a reductive tricarboxylic acid cycle (11). Their capacity to use organic carbon sources is usually limited to acetate, pyruvate, and propionate. Their photosynthetic apparatus consists of the cytoplasmic membrane (containing the reaction center bacteriochlorophyll a and the electron transport chain) and special light harvesting (antenna) organelles, the chlorosomes (containing bacteriochlorophyll c, d, or e and carotenoids).

A similar fine structure and pigment arrangement is found in the Chloroflexaceae (flexible, filamentous, gliding). The species of this family, however, are able to grow photoheterotrophically at the expense of organic carbon sources (38). The Chromatiaceae (classical purple sulfur bacteria) are partly obligate, partly facultative photolithotrophs. Only the latter utilize a wide variety of organic carbon compounds including sugars, fatty acids, and amino acids. This group, therefore, is physiologically rather similar to the Rhodospirillaceae (classical purple non-sulfur bacteria) (37). In both families, the Chromatiaceae and the Rhodospirillaceae, CO_2 is fixed via the Calvin cycle; the light harvesting pigments (bacteriochlorophyll a or b, carotenoids) and the reaction center bacteriochlorophyll a (or b) (plus the electron transport chain) are located in the cytoplasmic membrane that is enlarged many times by specific infoldings into the cytoplasm (intracytoplasmic membranes, present as vesicles, lamellar stacks, tubes, or polar cups). All species of the Rhodospirillaceae are photoheterotrophs, but many of them are also able to grow photolithotrophically (with H_2, H_2S, or thiosulfate), i.e., like the Chromatiaceae (35). So far, among the cyanobacteria (Cyanophyceae, blue green algae) 11 species have been described that are capable of living in anoxic environments by anoxygenic photosynthesis, using H_2S instead of H_2O as the electron donor and producing elemental sulfur (7,8,10,15,18). In cyanobacteria, CO_2 is fixed via the Calvin cycle. Thiobacillus denitrificans is the only facultative anaerobe of the genus Thiobacillus; it can perform a nitrate respiration (forming N_2) coupled to the oxidation of reduced sulfur compounds and fixes CO_2 via the Calvin cycle.

The oxidation of sulfide yields either sulfate or elemental sulfur. The latter is temporarily stored within the cells (most Chromatiaceae) or deposited outside (Chlorobiaceae, Chloroflexaceae, Ectothiorhodospira species). If sulfur is the end product (cyanobacteria, several Rhodospirillaceae species), it is deposited outside the cells.

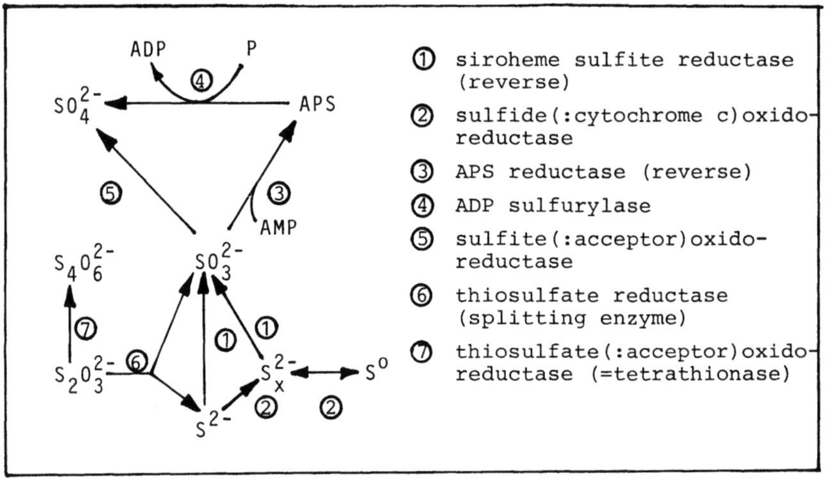

FIG. 1 - Enzymatic steps of anoxic sulfur oxidation in phototrophic and chemolithotrophic bacteria. Abbreviations: AMP, adenosine monophosphate; ADP, adenosine diphosphate; APS, adenylyl sulfate; P, inorganic phosphate.

The enzymatic steps listed in Fig. 1 are involved in anoxic sulfur oxidation in the following bacteria (52): (1) Reverse siroheme-containing sulfite reductase in Chromatium vinosum (44) and Thiobacillus denitrificans (42). (2) Sulfide (:cytochrome c) oxidoreductase in Chlorobium spec. and Chromatium vinosum; also nonenzymatic cytochrome c reduction occurs (52). (3) Adenylylsulfate reductase is present in Thiobacillus denitrificans, Thiocapsa roseopersicina, Chromatium vinosum,

Chlorobium limicola, Thiocystis violacea, but not in Ectothiorhodospira or any of the Rhodospirillaceae (4,53). (4) ADP sulfurylase occurs in all species that contain APS reductase, not in those that lack APS reductase (53). (5) Sulfite (:acceptor) oxidoreductase, independent of AMP, has been found in Chromatium vinosum (heme enzyme), Thiocapsa roseopersicina, Ectothiorhodospira species, and Rhodopseudomonas sulfidophila (53). (6) Thiosulfate reductase (or other thiosulfate-splitting enzymes including rhodanase) occurs in almost all phototrophic bacteria studied so far, as well as in Thiobacillus denitrificans (43,53). (7) The tetrathionate-forming thiosulfate: cytochrome c oxidoreductase was found in Chlorobium limicola f. thiosulfatophilum and in Rhodopseudomonas palustris. In Chromatium vinosum the better electron acceptor of this enzyme appears to be a high potential non-heme iron protein (14).

In spite of our limited knowledge in this field, it is already clear that parallel pathways exist in phototrophic anoxic sulfur oxidation, and that the pathway of sulfur in Thiobacillus denitrificans appears to be rather similar to that in Chromatium vinosum.

Oxic (Aerobic) Microbial Oxidation of Sulfur Compounds

In the presence of oxygen, bacteria oxidizing reduced sulfur compounds have to compete with oxygen itself, as nonbiological oxidation by oxygen is a typical property of the "unstable" sulfide, sulfite, thiosulfate, and polythionate ions. This enormously complicates the study of sulfur metabolism in aerobic sulfur-oxidizing bacteria. Although the instability of reduced sulfur species is commonly assumed, thiosulfate, trithionate, and tetrathionate are reasonably stable in pure solutions in the laboratory, as long as transition metal ions or sulfite and/or sulfide are not present to cause rapid disproportionation.

The only well studied pure cultures are those of the genera Thiobacillus and Thiomicrospira, both small cell gram-negative bacteria. Proper cultivation of Beggiatoaceae, Achromatiaceae, and the genera Thiobacterium, Macromonas, Thiovulum, etc. (all with large, conspicuous cells), is impeded by their simultaneous demand for low sulfide and low oxygen concentrations. There are very few pure cultures of Beggiatoa. Sulfolobus is a chemolithotrophic sulfur-oxidizing bacterium of extraordinary properties. It belongs to the archaebacteria (57), is an extreme acidothermophile, and is adapted to life in hot sulfur springs (5).

The enzymatic steps shown in Fig. 1 have been found to occur in thiobacilli, although not all of them in every species. Thus, APS reductase ③ and siroheme sulfite reductase ① occur only in T. denitrificans and in T. thioparus ((33,43) and Isamu Suzuki, personal communication). (Sirohemes are iron tetrahydroporphyrins of the isobacteriochlorin type with eight carboxylate side chains. They occur as prosthetic groups in certain sulfite reductases and nitrite reductases. Their biosynthesis branches off the biosynthetic pathway of cytochromes at an early intermediate step) (48).

Sulfite: cytochrome c oxidoreductase ⑤ is present in T. thioparus, T. novellus, T. neapolitanus, T. thiooxidans, T. ferrooxidans (41). A thiosulfate-splitting enzyme ⑥ occurs in T. novellus, T. intermedius, T. neapolitanus, tetrathionase ⑦ in T. thioparus, T. ferrooxidans, and T. neapolitanus (41).

Sulfur metabolism of aerobic thiobacilli involves the participation of molecular oxygen in enzymatic (51) and non-enzymatic steps as well as purely chemical reactions between different sulfur compounds (41,48,51).

Recent work by Kämpf and Pfennig (24) has shown that many species of the Chromatiaceae are also able to grow chemolithoautotrophically at the expense of sulfide and thiosulfate in the dark at low oxygen tension (e.g., Thiocapsa roseopersicina, Thiocystis violacea, Chromatium vinosum, and others). This property is not shared by the Chlorobiaceae and the large cell Chromatium species.

In addition to chemolithoautotrophic sulfur-oxidizing bacteria there are numerous species of heterotrophic bacteria, which are able to oxidize sulfide and thiosulfate, e.g., in/or above anoxic waters of marine basins (54).

Dissimilatory Reduction of Sulfur Compounds

Dissimilatory sulfate-reducing bacteria are strict anaerobes and use organic carbon compounds or H_2 as electron donors (39). Until recently, only Desulfovibrio, Desulfomonas, and Desulfotomaculum species that grew on lactate or ethanol had been cultivated. However, ecological work had shown that in marine (saline) environments, sulfate reduction accounts for about 50% of organic carbon mineralization (21) and had strongly suggested a wider physiological variety for these bacteria. Widdel (55) has now isolated and described a whole range of species and genera that are able to use acetate, formate, propionate, fatty acids up to C_{18}, and even benzoate. It seems of great importance for geochemistry that many of these newly detected organisms can dissimilate these carbon compounds completely to CO_2. These genera display a wide morphological variety expressed by their names: Desulfobulbus, Desulfobacter, Desulfococcus, Desulfonema, and Desulfosarcina. The fact that even gliding species (Desulfonema) and autotrophic forms (Desulfosarcina variabilis, Desulfonema limicola, growing with H_2, CO_2, and sulfate) exist, indicates that the sulfate-reducing bacteria may represent a phylogenetically rather diverse group. Even within the "classical" genera Desulfovibrio and Desulfotomaculum there exists a wide range of desoxyribonucleic acid base ratios indicative of great genetic diversity.

The question of (anaerobic) methane oxidation by dissimilatory sulfate-reducing bacteria has not yet been settled. It has been shown, however, that many methanogenic bacteria have the capacity to generate and to oxidize methane at the same time (59). Thus, considerable amounts of acetate may be formed that would allow growth of sulfate-reducing bacteria under natural conditions.

Sulfur metabolism has only so far been studied in Desulfovibrio and Desulfotomaculum species (Fig. 2).

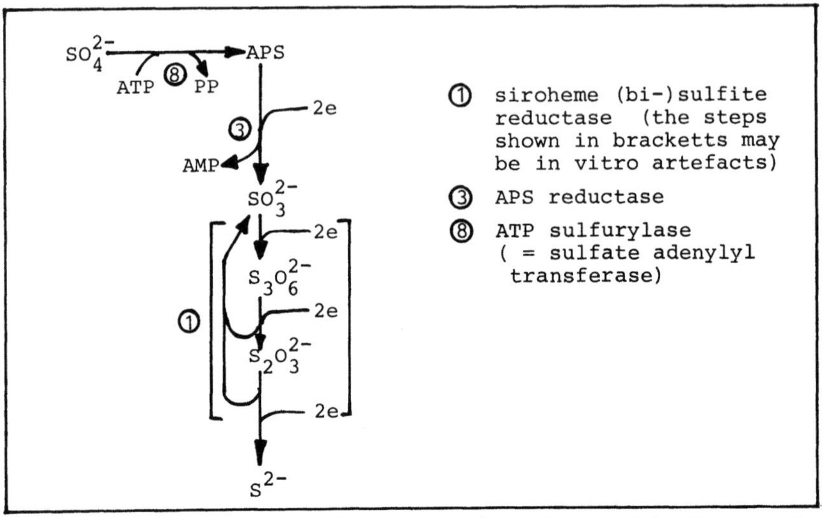

FIG. 2 - Dissimilatory sulfur metabolism in Desulfovibrio and Desulfotomaculum species (39).

The dissimilatory sulfate-reducing bacteria also use sulfite and thiosulfate readily as electron acceptors. In laboratory cultures, growth on these substrates is superior to that on sulfate.

Sulfite reduction to sulfide occurs in the strictly anaerobic bacterium Clostridium pasteurianum. The enzyme involved is an inducible "dissimilatory type" sulfite reductase (28). This bacterium is not capable of dissimilatory sulfate reduction.

Incidental reduction of elemental sulfur under anaerobic conditions is well known in many microorganisms, but in three groups of bacteria this process is physiologically necessary: a) The phototrophic purple and green bacteria that use intra- and extracellular sulfur as the electron acceptor for their (slow) fermentative (anaerobic) maintenance metabolism in the dark (52). b) Facultative dissimilatory sulfur-reducing bacteria represented by certain strains of sulfate-reducing bacteria that can reduce sulfur when sulfate is absent (2); also saprophytic Campylobacter strains (27,58). c) Obligate dissimilatory sulfur-reducing bacteria were isolated from marine muds and described as Desulfuromonas acetoxidans (36). These bacteria usually live in syntrophism with Chlorobiaceae species.

Assimilation of Sulfur Compounds

Many bacteria in anoxic environments are unable to assimilate sulfate (e.g., the Chlorobiaceae, certain members of the Chromatiaceae, and all methanogenic bacteria). All of these assimilate sulfide for the purpose of sulfur amino acid biosynthesis. Most purple bacteria can assimilate sulfide facultatively as well as sulfate. Sulfate assimilation is, however, an energetically more expensive process (i.e., requiring more ATP) than sulfide assimilation.

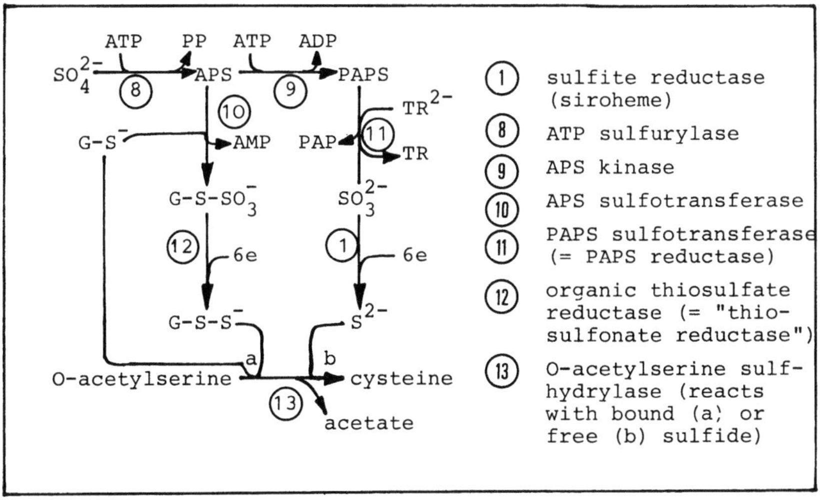

FIG. 3 - Enzymatic steps in assimilatory sulfur metabolism - the APS pathway and the PAPS pathway (34).
G-S$^-$, reduced glutathione; PAPS, 3'-phosphoadenylyl sulfate; PAP, 3'5'-adenosine bisphosphate; PP, pyrophosphate; TR, thioredoxin.

Fungi, plants, and most aerobic bacteria depend on assimilatory sulfate reduction. The pathway shows an interesting dichotomy ((10) versus (11)) that only occurs within the bacteria (including Rhodospirillaceae and cyanobacteria); while all green plants contain APS-sulfotransferase, fungi contain PAPS-sulfotransferase.

The Oxygen-free Sulfuretum

Under anoxic conditions, there is a "small" and a "large" sulfur cycle (Fig. 4, lower part). The small cycle (S^o/S^{2-}) maintained, for instance, by Chlorobiaceae and Desulfuromonas, requires only light, bicarbonate and acetate (or ethanol). Since sufficient elemental sulfur is present in marine sediments (22) near the redoxcline (20), this might be the way Chlorobiaceae survive in this habitat. Even a small amount of sulfur may guarantee a high turnover in a closed system (36).

The large cycle (SO_4^{2-}/ S^{2-}) is mediated by phototrophic sulfur bacteria and dissimilatory sulfate-reducing bacteria and requires light, bicarbonate, and certain organic compounds. Primary production by an anoxic sulfuretum occurs by phototrophic as well as by certain autotrophic sulfate-reducing bacteria, and in the presence of nitrate also by Thiobacillus denitrificans and physiologically similar forms. Chlorobiaceae are known to excrete 15-30% of the photosynthetically fixed carbon as organic acids (35), thus providing part of the nutrients for sulfate- and sulfur-reducing bacteria.

In an oxygen-free environment there is no strong redox gradient, although there are gradients imposed by gravity (sedimentation) and light. Chromatiaceae have higher light requirements than Chlorobiaceae. Therefore the latter usually occur (in sediments as well as in stratified lakes) below the former. In syntrophic associations with sulfur oxidizers, good growth of Chlorobiaceae may even be maintained at light intensities as low as 5-10 lux at which no growth of Chromatiaceae occurs (3). Anoxygenic primary production may account for up to 90% of the total primary production in stratified water bodies (9), although primary production of this type is usually smaller due to light limitations in the anoxic zone (31).

A proper sulfuretum in the sense of Baas-Becking (1) forms a more or less closed system in which sulfur is continuously cycled between the oxidized and reduced state. Therefore it seems unlikely that undisturbed sulfureta contributed to the accumulation of "biogenic" sulfide ores billions of years ago. The more closely ancient systems approached the state of a proper sulfuretum, the smaller the chance of detecting fossil remains in the stratigraphic record. Normal, undisturbed, present-day marine sediments do not contain sufficient heavy metal sulfides to be considered "sulfide ore deposits in statu nascendi," whereas they do contain nearly perfect sulfureta in the sense of Baas-Becking.

The Sulfuretum Under the Influence of Oxygen

Fully oxygenated sulfureta do not exist, because dissimilatory sulfate reducers are strict anaerobes. The presence of O_2 superimposes a redox gradient on the anoxic environment; as a consequence, strictly aerobic and anaerobic bacteria are separated, and their cooperation must be guaranteed by the diffusion of sulfur compounds and of other compounds across the redoxcline (cf. Fig. 4).

Furthermore, microaerobic bacteria and anaerobes that tolerate low oxygen tensions will appear stratified within the redox gradient according to their specific O_2 needs and tolerances. Cellular motility, chemotaxis, and phototaxis are ecologically advantageous in this setting. It has been found, however, that in a diurnal cycle, the same layer of an environment may fluctuate between oxic and anoxic conditions, while not all of the microorganisms respond by motility (23). Permanently multilaminated oxic and anoxic conditions have also been demonstrated (26). Therefore, mutual adaptation or at least tolerance to oxic/anoxic conditions should be relatively common to bacteria

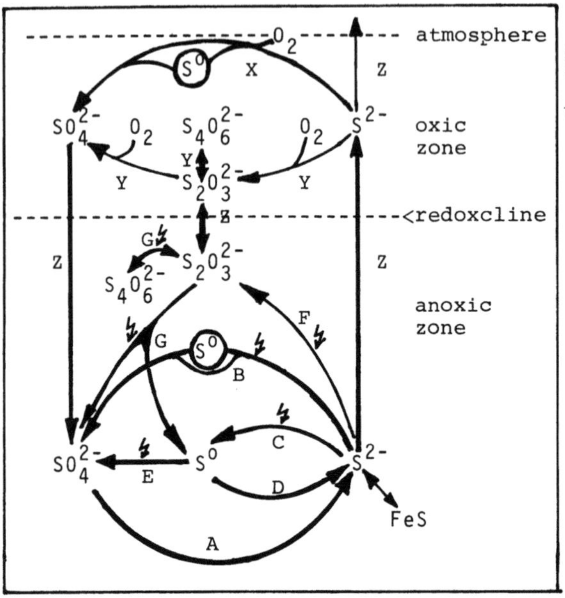

FIG. 4 - Sulfur transformations in the sulfuretum. Encircled S^0, intracellular sulfur; ϟ, light-driven; capital letters indicate the participation of the following bacteria: A, sulfate-reducing bacteria; B, Chromatiaceae (via intracellular sulfur), Rhodopseudomonas sulfidophila; C, Chlorobiaceae, Ectothiorhodospira, cyanobacteria, some Rhodospirillaceae; D, Desulfuromonas, some Desulfovibrio, Campylobacter, and Chromatiaceae and Chlorobiaceae in the dark; E, Chlorobiaceae, Chromatiaceae; F, some Chlorobiaceae; G, some Chromatiaceae, some Chlorobiaceae, some Rhodospirillaceae, Thiobacillus denitrificans; X, Beggiatoa and other large cell thiobacteria (via intracellular sulfur, thiobacilli, and Chromatiaceae in the dark; Y, thiobacilli, chemical oxidations; Z, diffusion.

in these environments. The oxic zone of the habitat contains the diverse forms of aerobic and microaerobic sulfide oxidizers (Beggiatoceae, thiobacilli, and large cell thiobacteria). The majority of the Chromatiaceae also participate in oxic sulfide oxidation in the dark.

THE EVOLUTION OF BACTERIA INVOLVED IN THE SULFUR CYCLE

During the last few years, several conservative macromolecular constituents of contemporary bacteria have been used as the basis for the construction of phylogenetic trees and dendograms: amino-acid sequences of cytochromes c and iron sulfur proteins (ferredoxins, etc.), tertiary structure of cytochromes c, DNA-DNA and DNA-RNA homologies, oligonucleotide patterns of ribosomal RNAs, peptide sequences of cell wall mureins, lipopolysaccharides of outer cell wall layers. Although most of these studies are of limited groups of bacteria, a fairly consistent picture is beginning to emerge. Surprisingly, this picture is in satisfactory agreement with the current hypothesis of biochemical and bioenergetic evolution brought forward by Peck (33), Broda (6), and others. The most exciting results have come from 16s RNA oligonucleotide studies by Carl Woese's group (13,56,57): Besides detection of a principal dichotomy within the prokaryotes (archaebacteria versus eubacteria) (13, 57), important information about phylogenetic relations between phototrophic bacteria and sulfate-reducing bacteria has been gained. Figure 5 presents a 16s RNA-based dendrogram that includes a number of the bacterial species which are involved in the sulfur cycle. So far, no 16s RNA data are available for chemolithotrophic sulfur bacteria and for the newly isolated sulfate-reducing bacteria of Widdel (55). The dendrogram shows that the Chloroflexaceae, the Chlorobiaceae, and the cyanobacteria each form rather old, phylogenetically isolated groups ((25,49); O. Kandler, personal communication), whose ancestors split off the bulk of the ancestral Gram-negative bacteria even before the ancestral Gram-positive bacteria had branched off (49).

With respect to the development of the bacterial sulfur cycle, it appears that the ancestors of the green phototrophic sulfur bacteria were the oldest sulfide oxidizers, closely followed by the ancestral cyanobacteria. The latter probably started as anaerobic sulfide oxidizers too, a capacity still retained by many of them. The present purple phototrophic bacteria (Chromatiaceae, Rhodospirillaceae) constitute a major phylogenetic group which is genealogically intermixed with many classically well-known nonphototrophic Gram-negative bacteria (16).

The ancestor of Desulfovibrio is one of the earlier progeny of this old purple bacterial group, it split off at S_{AB} = 0.3.

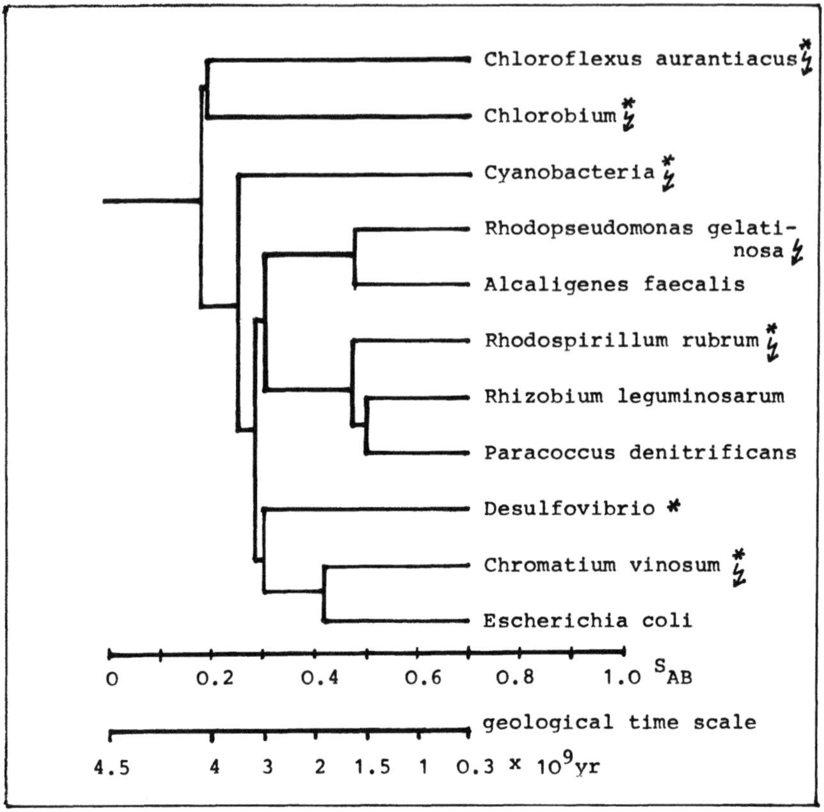

Fig. 5 - Dendrogram for some Gram-negative bacteria, based on 16s RNA oligonucleotide mapping (O. Kandler, personal communication). S_{AB}: similarity coefficient (1.0 = identity; 0.0 = no relatedness). * , involved in mass turnover of the bacterial sulfur cycle; ↯ , phototrophic.

To my knowledge, O. Kandler was the first to correlate S_{AB} values with the geological time scale. In a recent lecture, he proposed that the beginning of iron oxidation (continental red beds) 1.5 - 2 · 10^9 yr ago might correspond with the split of anaerobic or phototrophic groups into present aerobic or microaerobic orders and families at S_{AB} values of 0.4 - 0.5.

The probable beginning of noticeable sulfur isotope fractionation about 2.8 - 3.1 · 10^9 yr ago (45) might be equated with the emergence of the ancestral Desulfovibrio at S_{AB} = 0.3.

Before 2.8 · 10^9 yr ago, sulfur isotope fractionation data disclose relatively small fractionations; this might be due to the predominance of unstable sulfur species (such as elemental sulfur and thiosulfate), the reduction of which did not result in fractionation as intense as fractionation associated with sulfate reduction.

One could speculate that phototrophic bacteria with the ability of intracellular sulfur storage (e.g., of the Chromatium and Thiocapsa type) have an advantage over those which deposit elemental sulfur outside the cells (e.g., of the Chlorobium or Ectothiorhodospira type), where it is readily available to sulfur-reducing bacteria. Therefore, the appearance of intracellular sulfur storage might have been a prerequisite for phototrophic sulfur oxidation to sulfate and hence accumulation of sulfate. On the other hand, the appearance of O_2 in solutions of reduced sulfur compounds would also have led to sulfate accumulation.

In the geochemical sense, the biochemical reactions carried out by organisms rather than the organisms themselves are the important factors. A particular reaction, or set of reactions, e.g., sulfate reduction, is the expression of a very small part of the entire genome.

Prior to the appearance of dissimilatory sulfate reduction, a real sulfuretum (by definition) did not exist, as no sulfate could be recycled to sulfide. The invention of sulfate activation (ATP sulfurylase), a step that requires two energy-rich phosphate bonds, was the decisive step. The reaction in the opposite direction (ADP sulfurylase or perhaps already ATP sulfurylase), however, is apparently older. Furthermore, dissimilatory sulfur or thiosulfate reduction could have preceded sulfate reduction. Both pathways had been preformed by ancestors of the Chlorobiaceae, though in the opposite direction. Hypothetically, even sulfite or trithionate reduction could have been precursors of sulfate reduction, taking into account the peculiar in vitro behavior of dissimilatory sulfite reductase (cf. Fig. 2).

Postgate (39) lists a number of characters of Desulfovibrio that he considers to be "primitive" compared to other bacteria:

1) Possession of hydrogenase linked to phosphoroclastic pyruvate breakdown.
2) Use of ferredoxin and rubredoxin in electron transport.
3) Reductive carboxylation of acetate to form pyruvate (a primitive CO_2 assimilation process).
4) Mixotrophic assimilation of acetate and CO_2 under H_2.
5) Sirohydrochlorin in Desulfovibrio siroheme sulfite reductase has been considered "primitive" in comparison with other porphyrin molecules, because it is formed in the biosynthetic pathway from a relatively early porphyrin precursor.

   On the other hand, cytochromes (formed at a later step in porphyrin biosynthesis) would then have to be considered as "advanced" molecules. The fact that these are also generally found in sulfate-reducing bacteria demonstrates that the possession of certain "primitive" traits does not imply that an organism as a whole is "primitive."

All of the traits listed above may be considered as part of the phototrophic bacterial heritage.

Dissimilatory sulfate reduction is considered older than assimilatory sulfate reduction (41,47,48). A real necessity for assimilatory sulfate reduction arose when the availability of reduced sulfur became limited due to increasing oxygen production in the biosphere. Carrier-bound intermediates might have become necessary in assimilatory sulfate reduction after the anoxic environment became merely an ecological niche after the development of water-splitting photosynthesis, because they (glutathione-S-sulfonate and -persulfide) are not autoxidizable as are sulfite and sulfide. Later the carrier-bound PAPS pathway may have evolved from the carrier-bound APS pathway. The PAPS pathway is present in bacteria (including part of the cyanobacteria) and fungi, the APS pathway in other bacteria (including the other part of the cyanobacteria), and in green plants.

It could be argued, on the other hand, that a requirement for reduced sulfur for cell biosynthesis is a more universal phenomenon than dissimilatory sulfate reduction, and that at a time of considerable dissimilatory reduced sulfur supply this requirement could have provided a selective pressure towards assimilation rather than dissimilation of oxidized sulfur. Sirohydrochlorin (siroheme) is also a member of the prosthetic group of assimilatory sulfite reductases.

A major question in evolution is still: "when did oxygenic photosynthesis develop?" The carbon isotope fractionation attributed to photosynthesis in the $3.76 \cdot 10^9$ yr old sediments of Isua, West Greenland (46), might indicate the very beginning of measurable CO_2 fixation via the Calvin cycle by photosynthesis, which would correspond to an S_{AB} value in the range of 0.2 in Fig. 5, i.e., the divergence of the cyanobacterial and purple-bacterial ancestral branch from the ancestral green bacteria. This does not necessarily implicate development of oxygenic photosynthesis at that time. The ancestral cyanobacteria could well have existed by anoxygenic photosynthesis using only photosystem I, driven by sulfide. Relatively high tolerance to sulfide has also been shown by strains that can only grow by oxygenic photosynthesis (15,50). H_2S prevents an accumulation of O_2 in the cyanobacterial environment (50) and could very well have done so for about two billion years. As in some cyanobacteria sulfite, cysteine, and hydrazine act as electron donors for photosystem II (34), thus permitting anoxygenic photosynthesis dependent on both photosystems; the suggestion has been made that the two photosystems evolved in cyanobacteria before the invention of water-splitting (34).

The earliest water-splitting photosynthetic bacteria perhaps still depended on an external source of reduced sulfur for assimilatory purposes. This might have led to a steady-state that was balanced by photosynthetic oxygen production on the one hand and by the destruction of sulfide reservoirs by oxygen on the other hand, thus controlling the growth rate of such organisms. The supply with additional reduced sulfur from deep in the earth's crust could have disturbed this balance episodically. The system may have produced a large reservoir of sulfate without substantially raising the concentration of molecular oxygen. In addition, as long as the oxygen level in the atmosphere was low, the photochemical processes that are now restricted to the high atmosphere could have been active near the earth's surface and intensified the reaction of O_2 with oxidizable matter. Thus a minor increase in the oxygen level must have had a dramatic effect on the remaining reservoirs of available sulfide.

With the invention of assimilatory sulfate reduction, water-splitting phototrophic bacteria (the ancestral cyanobacteria) could free themselves from this steady-state. Therefore, assimilatory sulfate reduction might have had an enormous influence on the development of higher atmospheric oxygen levels. These ideas have been developed by Heimo Nielsen (personal communication) and deserve special attention with respect to the entire O_2 level discussion.

Filamentous bacteria with gliding motility are adapted to live in or on particulate solid environments such as sediments. Such forms exist among the cyanobacteria (e.g., Oscillatoria), the Chloroflexaceae (e.g., Chloroflexus, Oscillochloris), the Beggiatoaceae (e.g., Beggiatoa, Thiothrix), the sulfate-reducing bacteria (Desulfonema), and perhaps even in the purple bacteria (R. Castenholz, personal communication). So far there are not sufficient chemotaxonomic data to decide whether this is of phylogenetic importance. The cyanobacteria with their wide morphological variety have been shown to form a phylogenetically well defined cluster (as fas as they have been studied to date). Largely for reasons of morphology, Pringsheim (40) considered the Beggiatoaceae and other large cell thiobacteria to be "colorless blue-greens." Flagella are of great advantage only in aqueous environments (including interstitial water) and do not exist in present Chlorobiaceae (all nonmotile), Chloroflexaceae, or cyanobacteria. They may have been invented after the divergence of the cyanobacteria.

In the following I shall give a summary of my present, rather speculative, opinions concerning the evolution of the bacteria involved in the sulfur turnover of the biosphere:

1. The first bacteria derived energy from the fermentation of "Oparin broth." They used sulfide as assimilatory sulfur source. This trait is still found in many strict anaerobes. Exhaustion of free organic matter yielded an evolutionary pressure towards utilization of light as an energy source and CO_2 as a carbon source.
2. The earlier real phototrophic bacteria had one light reaction (mediated by bacteriochlorophyll a) and an electron transport chain that included ferredoxins, cytochromes, etc. They fixed CO_2 via a reductive tricarboxylic acid cycle. The electron donor for photosynthesis was H_2S, which was oxidized to extracellular S^o. The overall fractionation of sulfur and carbon isotopes was very small. In the dark, these bacteria used S^o as an electron sink in a fermentative maintenance metabolism.
3. Metabolism using S^o as an electron sink became the main energy source for certain bacteria that lost their ability to photosynthesize and became ancestors of the Desulfuromonas-like bacteria, i.e., the dissimilatory sulfur reducers. These - together with the ancestral phototrophic sulfide oxidizers - formed the first (small cycle) sulfuretum.
4. The early phototrophs then acquired the ability to oxidize S^o to sulfate, perhaps stepwise via sulfite, by siroheme sulfite reductase and APS reductase. Their dark fermentative metabolism did not develop further.
5. One group (the ancestors of the contemporary Chloroflexaceae and Chlorobiaceae) split off and developed special light harvesting devices (chlorosomes) with special new antenna bacteriochlorophylls that allowed them to live at low light intensities. The ancestral Chloroflexaceae kept gliding motility and - much later - became facultative aerobes, although they retained sulfide photosynthesis. Their conspicuous filamentous and trichome morphology may have given rise to early stromatolites. Nonphototrophic flexibacteria and perhaps other gliding bacteria may have developed from them.

The ancestral Chlorobiaceae stayed non-motile, depending on sulfide-photosynthesis ($HS^- \longrightarrow S^O \longrightarrow SO_4^{2-}$, after inventing APS reductase). They did not develop sulfate assimilation. The present forms are all strict anaerobes and strict photolithotrophs and fix CO_2 via the reductive tricarboxylic acid cycle.

6. The remaining group of phototrophs developed the reductive pentose phosphate (Calvin) cycle. They were the ancestors of the cyanobacteria, of the purple bacteria, and perhaps of most other eubacteria (13).
7. The cyanobacteria diverged from these. They lost APS reductase, changed to the shorter wavelength-absorbing chlorophyll a, and added phycobilisomes containing phycobilins as light harvesting devices. Furthermore, they added a second light reaction that finally enabled them to split water and evolve O_2. From them the red algal chloroplasts, the earlier Prochlorales (prokaryotes with chlorophylls a and b and oxygenic photosynthesis that were the likely ancestors of plant chloroplasts (30)), the Beggiatoa-like and large cell thiobacteria probably had their origin.
8. The organisms of the stem remaining after the divergence of the cyanobacteria were still phototrophic, contained bacteriochlorophyll a, oxidized sulfide to sulfate, and had APS reductase. They developed flagella and split into at least three large groups (16), each containing phototrophs (now classified in the Chromatiaceae and Rhodospirillaceae) and nonphototrophs. The Rhodospirillaceae and part of the Chromatiaceae either lost APS reductase completely or replaced it by sulfite oxidoreductase; some of the Rhodospirillaceae lost the ability to use reduced sulfur compounds as electron donors at all.
9. The dissimilatory sulfate-reducing bacteria developed from phototrophic sulfur bacteria, e.g., of the ancestral Chromatiaceae type. First, the latter bacteria may have changed their dark fermentative metabolism from elemental sulfur reduction (sulfur is but a two-electron sink) to sulfate reduction (sulfate can take up eight electrons; there are recent textbooks that prefer to call dissimilatory sulfate

reduction "sulfide fermentation" (!) (17)), using enzymes they possessed for oxidative (light) sulfur metabolism (APS reductase, siroheme sulfite reductase), and at the same time maintaining their phototrophic mode of life during daytime. In the same way in which the photosynthetic electron transport chain (cytochromes, quinones, flavins, etc.) later became involved in O_2 respiration, it may have also become involved in dissimilatory sulfate reduction. The next step was the loss of photosynthesis and of the Calvin cycle, probably due to better survival in deeper sediment layers with higher organic nutrient content, but permanently dark. This step finally led to the present-day sulfate-reducing bacteria.

A study of the CO_2-fixing mechanism of the autotrophic desulfobacterial genera will have to show whether they possess the Calvin cycle or a reverse Krebs cycle. The development from phototrophy to dissimilatory sulfate reduction may well have happened, like the change to O_2 respiration, along different lines of the Gram-negative stem. If so, this would help to explain the morphological and physiological variety of present-day sulfate-reducing species.

10. Like other Gram-negative bacteria, the thiobacilli were derived from the three large groups of the purple bacterial stem. They lost the ability to photosynthesize, but retained Calvin cycle and sulfur metabolism. Some retained APS reductase and siroheme sulfite reductase, others replaced these enzymes by, e.g., sulfite: cytochrome c oxidoreductase and O_2-involving mechanisms. The present genus Thiobacillus contains several intermediate metabolic types between strict autotrophic and heterotrophic forms.

Acknowledgement. I thank P. Trudinger, H. Nielsen, Y. Cohen, I. Kaplan, T. Brock, and R. Hallberg for their critical and very helpful remarks and advice in the preparation of the final manuscript.

REFERENCES

(1) Baas-Becking, L.G.M. 1925. Studies on the sulphur bacteria. Ann. Bot. 39: 613-650.

(2) Biebl, H., and Pfennig, N. 1977. Growth of sulfate-reducing bacteria with sulfur as electron acceptor. Arch. Microbiol. 112: 115-117.

(3) Biebl, H., and Pfennig, N. 1978. Growth yields of green sulfur bacteria in mixed cultures with sulfur and sulfate reducing bacteria. Arch. Microbiol. 117: 9-16.

(4) Bowen, T.J.; Happold, F.C.; and Taylor, B.F. 1966. Studies on adenosine-5'-phosphosulphate reductase from Thiobacillus denitrificans. Biochim. Biophys. Acta 118: 566-576.

(5) Brock, T.D. 1978. Thermophilic Microorganisms and Life at High Temperatures. New York: Springer Verlag.

(6) Broda, E. 1975. The Evolution of the Bioenergetic Processes. Oxford: Pergamon Press.

(7) Castenholz, R.W. 1976. The effect of sulfide on the blue-green algae of hot springs. I. New Zealand and Iceland. J. J. Phycol. 12: 54-68.

(8) Castenholz, R.W. 1977. The effect of sulfide on the blue-green algae of hot springs. II. Yellowstone National Park. Microbial. Ecol. 3: 79-105.

(9) Cohen, Y.; Krumbein, W.E.; and Shilo, M. 1977. Solar Lake (Sinai) 2. Distribution of photosynthetic microorganisms and primary production. Limnol. Oceanogr. 22: 609-620.

(10) Cohen, Y.; Paden, E.; and Shilo, M. 1975. Facultative anoxygenic photosynthesis in the cyanobacterium Oscillatoria limnetica. J. Bact. 123: 855-861.

(11) Evans, M.C.W.; Buchanan, B.B.; and Arnon, D.I. 1966. A new ferredoxin-dependent carbon reduction cycle in a photosynthetic bacterium. Proc. Natl. Acad. Sci. USA 55: 928-934.

(12) Fenchel, T.M., and Riedl, R.J. 1970. The sulfide system: a new biotic community underneath the oxidized layer of marine sand bottoms. Mar. Biol. 7: 255-268.

(13) Fox, G.E.; Stackebrandt, E.; Hespell, R.B.; Gibson, J.; Maniloff, J.; Dyer, T.A.; Wolfe, R.S.; Balch, W.E.; Tanner, R.S.; Magrum, L.J.; Zablen, L.B.; Blakemore, R.; Gupta, R.; Bonen, L.; Lewis, B.J.; Stahl, D.A.; Luehrsen K.R.; Chen, K.N.; and Woese, C.R. 1980. The phylogeny of prokaryotes. Science 209: 457-463.

(14) Fukumori, Y., and Yamanaka, T. 1979. A high-potential nonheme iron protein (HiPIP)-linked, thiosulfate-oxidizing enzyme derived from Chromatium vinosum. Curr. Microbiol. 3: 117-120.

(15) Garlick, S.; Oren, A.; and Padan, R. 1977. Occurrence of facultative anoxygenic photosynthesis among filamentous and unicellular cyanobacteria. J. Bact. 129: 623-629.

(16) Gibson, J.; Stackebrandt, E.; Zablen, L.B.; Gupta, R.; and Woese, C.R. 1979. A phylogenetic analysis of the purple photosynthetic bacteria. Curr. Microbiol. 3: 59-64.

(17) Gottschalk, G. 1979. Bacterial Metabolism. New York/ Heidelberg/Berlin: Springer Verlag.

(18) Imhoff, J.F.; Hashwa, F.; and Trüper, H.G. 1978. Isolation of extremely halophilic phototrophic bacteria from the alkaline Wadi Natrun, Egypt. Arch. Hydrobiol. 84: 381-388.

(19) Imhoff, J.F.; Sahl, H.G.; Soliman, G.S.H.; and Trüper, H.G. 1979. The Wadi Natrun: chemical composition and microbial mass developments in alkaline brines of eutrophic desert lakes. Geomicrobiol. J. 1: 219-234.

(20) Hallberg, R.O. 1972. Sedimentary sulfide mineral formation - an energy circuit system approach. Mineral. Deposita (Berl.) 7: 189-201.

(21) Jørgensen, B.B., and Cohen, Y. 1977. Solar lake (Sinai) 5. The sulfur cycle of benthic cyanobacterial mats. Limnol. Oceanogr. 22: 657-666.

(22) Jørgensen, B.B., and Fenchel, T. 1974. The sulfur cycle of a marine sediment model system. Mar. Biol. 24: 189-201.

(23) Jørgensen, B.B.; Revsbech, N.P.; Blackburn, T.H.; and Cohen, Y. 1979. Diurnal cycle of oxygen and sulfide microgradients and microbial photosynthesis in a cyanobacterial mat sediment. Appl. Environ. Microbiol. 38: 46-58.

(24) Kämpf, C., and Pfennig, N. 1980. Capacity of Chromatiaceae for chemotrophic growth. Specific respiration rates of Thiocystis violacea and Chromatium vinosum. Arch. Microbiol 127: 125-137.

(25) Kandler, O., and Schleifer, K.H. 1980. Systematics of bacteria. Fortschr. der Botanik 42: 234-252.

(26) Krumbein, W.E.; Buchholz, H.; Franke, P.; Giani, D.; Giele, C.; and Wonneberger, K. 1979. O_2 and H_2S coexistence in stromatolites. A model for the origin of mineralogical lamination in stromatolites and banded iron formations. Naturwissenschaften 66: 381-389.

(27) Laanbroek, H.J.; Stal, L.J.; and Veldkamp. H. 1978. Utilization of hydrogen and formate by Campylobacter spec. under aerobic and anaerobic conditions. Arch. Microbiol. 119: 99-102.

(28) Laishley, E.J., and Krouse, H.R. 1978. Stable isotope fractionation by Clostridium pasteurianum. 2. Can. J. Microbiol. 24: 716-724.

(29) Lauterborn, R. 1915. Die sapropelische Lebewelt. Ein Beitrag zur Biologie des Faulschlamms natürlicher Gewässer. Verh. Naturhist. Med. Ver. Heidelberg, N.F. 13: 395-481.

(30) Lewin, R.A. 1977. Prochloron, type genus of the Prochlorophyta. Phycologia 16: 217.

(31) Parkin, T.B., and Brock, T.D. 1980. Photosynthetic bacterial production in lakes: The effects of light intensity. Limnol. Oceanogr. 25: 711-718.

(32) Peck, H.D. 1966/67. Some evolutionary aspects of inorganic sulfur metabolism. Lectures on theoretical and applied aspects of modern microbiology, University of Maryland, College Park, MD.

(33) Peck, H.D.; Deacon, T.E.; and Davidson, I.T. 1965. Studies on adenosine 5'-phosphosulfate reductase from Desulfovibrio desulfuricans and Thiobacillus thioparus. Biochim. Biophys. Acta 96: 429-446.

(34) Peschek, G.A. 1978. Reduced sulfur and nitrogen compounds and molecular hydrogen as electron donors for anaerobic CO_2 photoreduction in Anacystis nidulans. Arch Microbiol. 119: 313-322.

(35) Pfennig, N. 1978. General physiology and ecology of photosynthetic bacteria. In The Photosynthetic Bacteria, eds. R.K. Clayton and W.R. Sistrom, pp. 3-18. New York and London: Plenum Press.

(36) Pfennig, N., and Biebl, H. 1976. Desulfuromonas acetoxidans gen. nov. and sp. nov., a new anaerobic, sulfur-reducing, acetate-oxidizing bacterium. Arch. Microbiol. 110: 3-12.

(37) Pfennig, N., and Trüper, H.G. 1977. The Rhodospirillales (phototrophic or photosynthetic bacteria). In CRC Handbook of Microbiology, eds. A.I. Laskin and H.A. Lechevalier, 2nd ed., vol. 1, pp. 119-130. Cleveland/OH: CRC Press.

(38) Pierson, B.K., and Castenholz, R.W. 1974. A phototrophic gliding filamentous bacterium of hot springs, Chloroflexus aurantiacus, gen. and sp. nov. Arch. Microbiol. 100: 5-24.

(39) Postgate, J.R. 1979. The Sulphate-reducing Bacteria. Cambridge: University Press.

(40) Pringsheim, E.G. 1963. Farblose Algen. Stuttgart: Fischer-Verlag.

(41) Roy, A.B., and Trudinger, P.A. 1970. The biochemistry of inorganic compounds of sulphur. Cambridge: University Press.

(42) Schedel, M., and Trüper, H.G. 1979. Purification of Thiobacillus denitrificans siroheme sulfite reductase and investigation of some molecular and catalytic properties. Biochim. Biophys. Acta 568: 454-467.

(43) Schedel, M. and Trüper, H.G. 1980. Anaerobic oxidation of thiosulfate and elemental sulfur in Thiobacillus denitrificans. Arch. Microbiol. 124: 205-210.

(44) Schedel, M.; Vanselow, M.; Trüper, H.G. 1979. Siroheme sulfite reductase isolated from Chromatium vinosum. Purification and investigation of some of its molecular and catalytic properties. Arch. Microbiol. 121: 29-36.

(45) Schidlowski, M. 1979. Antiquity and evolutionary status of bacterial sulfate reduction: sulfur isotope evidence. Origin of Life 9: 299-311.

(46) Schidlowski, M.; Appel, P.W.U.; Eichmann, R.; and Junge, C.E. 1979. Carbon isotope geochemistry of the 3.7 x 10^9 yr-old Isua sediment, West Greenland: implication for the archaean carbon and oxygen cycles. Geochim. Cosmochim. Acta 43: 189-199.

(47) Schiff, J.A. 1980. Pathways of assimilatory sulphate reduction in plants and microorganisms. In Sulphur in Biology, Ciba Foundation Symposium 72 (new series), pp. 49-69. Amsterdam: Excerpta Medica.

(48) Siegel, L.M. 1975. Biochemistry of the sulfur cycle. In Metabolic Pathways, Metabolism of sulfur compounds, ed. D.M. Greenberg, 3rd ed., vol. 6, pp. 217-286. New York/San Francisco/London: Academic Press.

(49) Stackebrandt, E., and Woese, C.R. 1979. Primärstruktur der ribosomalen 16s RNS - ein Marker der Evolution der Prokaryonten. Forum Mikrobiologie 2: 183-190.

(50) Steward, W.D.P., and Pearson, H.W. 1970. Effects of aerobic and anaerobic conditions on growth and metabolisms of blue-green algae. Proc. R. Soc. Lond. B. 175: 293-311.

(51) Suzuki, I. 1965. Incorporation of atmospheric oxygen-18 into thiosulfate by the sulfur-oxidizing enzyme of Thiobacillus thiooxidans. Biochim. Biophys. Acta 110: 97-101.

(52) Trüper, H.G. 1978. Sulfur metabolism. In The Photosynthetic Bacteria, eds. R.K. Clayton and W.R. Sistrom, pp. 677-690. New York and London: Plenum Press.

(53) Trüper, H.G. 1981. Photolithotrophic sulfur oxidation. In Metabolism of inorganic nitrogen and sulfur compounds, eds. H. Bothe and A. Trebst. Heidelberg: Springer Verlag, in press.

(54) Tuttle, J.H., and Jannasch, H.W. 1973. Sulfide and thiosulfate-oxidizing bacteria in anoxic marine basins. Mar. Biol. 20: 64-70.

(55) Widdel, F. 1980. Anaerober Abbau von Fettsäuren and Benzoesäure durch neu isolierte Arten Sulfat-reduzierender Bakterien. Doctoral Thesis, Univeristy of Göttingen, F.R. Germany.

(56) Woese, C.R., and Fox, G.E. 1977. Phylogenetic structure of the prokaryotic domain: the primary kingdoms. Proc. Natl. Acad. Sci. USA 74: 5088-5090.

(57) Woese, C.R.; Magrum, L.J.; and Fox, G.E. 1978. Archaebacteria. J. Mol. Evol. 11: 245-252.

(58) Wolfe, R.S., and Pfennig, N. 1977. Reduction of sulfur by spirillum 5175 and syntrophism with Chlorobium. Appl. Environ. Microbiol. 33: 427-433.

(59) Zehnder, A.J.B., and Brock, T.D. 1979. Methane formation and methane oxidation by methanogenic bacteria. J. Bact. 137: 420-432.

Mineral Deposits and the Evolution of the Biosphere, eds. H.D. Holland and M. Schidlowski, pp. 31-50. Dahlem Konferenzen, 1982.
Berlin, Heidelberg, New York: Springer-Verlag.

The Organic Geochemistry of Benthic Microbial Ecosystems

K. L. H. Edmunds, S. C. Brassell, and G. Eglinton
Organic Geochemistry Unit, University of Bristol
School of Chemistry, Bristol BS8 1TS, England

Abstract. The biogeochemical processes operating in aquatic bottom sediments are poorly understood. The activity of the benthos and the importance of their role in the geological past are unclear. Organic geochemical data reflect the action of benthic microbial communities. The study of selected natural and contrived environments is urged as a means for understanding present-day processes and for the development of a framework for the paleoenvironmental evaluation of ancient sediments. Organic geochemistry should provide a powerful tool for the study of benthic microbial activity during the Phanerozoic.

INTRODUCTION

The earliest-known forms of life are preserved in the Precambrian. These fossils consist of the remains of prokaryotic communities, stromatolites, which are believed to be the ancient counterparts of modern microbial mats. However, the fossil evidence is generally poor: although the macroscopic stromatolite structures are clearly discernible, the microfossils of bacteria themselves are less clear. In addition, valid organic geochemical data from these samples are difficult to obtain because of their low organic carbon content and because of post-depositional contamination. These difficulties in chemical analysis become less severe as we go to progressively younger sediments; in some sediments only chemical data survive where morphological remains have perished. Comparison of ancient environments with those of the present-day is the only way in which we can elucidate life in the past, and the evolution of the biosphere.

However, intrinsic problems exist. No complete species list for a microbial community and no complete chemical catalog for a species of microbe are known. Chemical diagenesis is only partially understood. The detailed interpretation of organic geochemical data is therefore difficult, but holds great promise for the future.

STATE OF KNOWLEDGE

The relationships between biota, organic compounds and metals in benthic ecosystems are poorly understood. Biota act as sources of organic matter, but the complexity of the food web is formidable. Preliminary work has shown that microorganisms-metal and organic-metal associations can be environmentally important, but their nature and extent are uncertain.

SEDIMENTARY ENVIRONMENTS

Biological marker compounds, particularly lipids, play a vital role in the study of environmental organic geochemistry. They provide information concerning the sediment inputs and the benthic microbiota but they do not yield a definitive record of biological activity in an environment. Throughout the system, selective processes, including microbial activity, destroy and alter organic matter and contribute new compounds. Visually, only a small proportion (ca. $< 5\%$) of the organic matter generated in the photic zone reaches the sediment. Thus, cycling processes involving organic and inorganic carbon are important within the water column, within sediments, and between the two. These processes (Fig. 1) are of major importance in the turnover of organic matter and minerals in aquatic environments.

The biogeochemistry of most types of sedimentary environments is poorly characterized. Cyclic and random fluctuations of conditions are presumably paralleled by variations in the nature of microbial populations. The degree of oxicity or anoxicity of sediments is of paramount significance: it determines whether aerobic or anaerobic bacterial communities dominate although both are usually present. Anaerobic bacteria are frequently abundant in aerobic sediments, the anaerobic colonies being encapsulated in microenvironments which "float" in aerobic zones and use metabolites created by the aerobic community. The preservation of organic matter is favored by reducing conditions and many organic-rich sediments are anoxic; this is true, for example, for sediments in the Black Sea and in the Cariaco Trench (6). Microenvironments may be important in bottom sediments. Oxygen and H_2S can coexist in the alternating, fine laminae of microbial mats inhabitated by populations of oxygenic and anoxygenic photosynthetic microbes (14). Certain microbial activities, for example, manganese oxidation are enhanced by association with surfaces (18). Nutrients such as NH_4^+ and NO_2^- are frequently concentrated in sediment pore waters (21) and permit high microbial activity within sediments.

The organic geochemistry of bottom sediments, particularly the composition of lipids, reflects variations in depositional environments. In oxic environments recycling of organic carbon is generally rapid, and lipid preservation is poor. It is often possible to infer the nature of the depositional environment from the lipids in sediments, but relatively little detailed information on the lipid input from specific species is available at present.

Simple Systems

The environments of bottom sediments and the associated benthic microbial communities are complex. Certain poorly characterized environments (such as the deep sea floor) may have a limited but specialized microbiota. Results from the study of relatively simple systems may help in the interpretation of more complex environments. For example, anoxic sediments are to a large extent free of the complications that arise from the activity of animals and plants. Though each environment is unique, a number of common factors do exist. Lipid signatures for certain compound classes (e.g., hopanoids) are remarkably similar in many types of sediments and in many areas. Biogeochemical features common to many sediments may reflect the operation of a common biota and/or of common biochemical processes. From the organic geochemical viewpoint, desirable, simple, "model" environments have: (a) a well characterized biota, (b) a well defined input, (c) freedom from anthropogenic pollution, (d) geological relevance, and (e) accessibility.

Microbial mats can satisfy these criteria. A number of studies have been carried out in this environment, including the evaluation of lipid composition (3), ecology (12), and radiolabelled lipid diagenesis (13). In addition, the microbes themselves have recently been examined in some detail; these studies have revealed a staggering diversity of prokaryotic species (17). Microbial mats are clearly more complex environments than had been believed, although they are not significantly complicated by the presence of eukaryotes. The geological relevance of present-day microbial mats to stromatolites as old as the Precambrian is well documented.

ACTIVITY OF BENTHOS

Bacteria interact intensely with their environment because they are small and hence have a large surface-to-volume ratio, and because they multiply rapidly.

Microbial activity in the deep sea is of particular interest because most of the ocean floor is at great depth. Bacteria have been estimated to constitute 10-20% of the total oceanic biomass. To date, the detailed study of deep sea benthic microbiota has been hampered by the technical difficulties of collecting and culturing bacteria without decompression. To our knowledge, only one obligate barophilic bacterium, a spirillum, has been isolated (31).

The role of large organisms in benthic ecosystems should not be overlooked. Time-lapse photography on the continental shelf and slope has revealed extensive activity by fish, sponges, starfish and burrowing worms at the sediment surface. Microbial activity also occurs and there is a great deal of symbiosis between microorganisms and large organisms. Indeed, the barophilic spirillum found by Yayanos et al. (31) was associated with deep sea amphipods. In general, the nature and extent of the associations between bacteria and eukaryotic organisms are poorly understood and their study is a challenging area of oceanic microbiological research.

Benthic microorganisms often obtain their food and energy from a variety of sources which are hierarchically related. For example, in microbial mats several prokaryotic species with complementary light requirements colonize appropriate horizons in the mat structure. In the deep sea, the microbial hierarchy will presumably reflect the generalized decreases with depth of such parameters as temperature, light intensity, dissolved oxygen, and nutrient concentration. The organic composition varies with depth because organic compounds are both removed and contributed by successive microbial communities. Hence, the diet of deeper living bacteria contains less of the conventionally "nutritious" organic compounds, such as sugars, compared to the less "nutritious" lipids. Indeed, many marine bacteria appear to be adapted to low nutrient levels (9). In situ feeding experiments suggest that deep sea bacteria have very slow metabolic rates (11). The substrates used, however, were "nutritious" substances such as sugars, and therefore may not have been representative of the normal local "diet."

Two metabolic processes unique to prokaroyotes: anaerobic respiration and chemolithotrophy (the ability to utilize inorganic compounds as sources of energy) may be of major significance in benthic ecosystems. At present the contribution of anaerobic bacteria to marine ecosystems is not well understood. It is known, however, that bacteria can catalyze redox reactions with very different electrode potentials (Fig. 1). A sequence of such reactions might occur in a sediment as it becomes progressively more anoxic. However, microorganisms are known to be able to adapt to fluctuating oxic/anoxic conditions (e.g., (4)). Sulfate reduction is well documented in marine environments and is a dominant reaction in anoxic marine basins with high organic inputs. The toxicity of the hydrogen sulfide produced during sulfate reduction leads to the nearly complete exclusion of organisms which are not resistant to sulfide.

In metal-rich sediments, insoluble sulfides (mainly FeS) are formed. Sulfate reduction is not confined to sediments; there are highly productive environments, for example, the Black Sea, where sulfate reduction occurs in the water column. Hence, the sulfur cycle may dominate the overall turnover of organic matter in particular settings.

The global significance of chemolithotrophy is difficult to assess. The sulfide oxidizers which colonize the area surrounding H_2S-emitting thermal vents (e.g., in the Galapagos (5)) are striking examples of chemolithotrophy. Bacteria in such settings form the basis of a well developed food web that includes communities of molluscs and other invertebrates. Bacterial populations vary with the type of environment. For example, archaebacteria (i.e., thermoacidophiles, halophiles, and methanogens) colonize environments (e.g., hot springs, hypersaline lakes) which are fatal for less tolerant microorganisms.

COMPOUNDS INDICATIVE OF MICROBIAL ACTIVITY: LIPIDS

Many microorganisms possess a characteristic suite of lipids (Table 1). In general, the products of biosynthetic pathways in bacteria appear to be less elaborate than those of higher organisms.

The distribution of lipids in sediments can often be used to determine their biological source or sources, even though

TABLE 1 - Bacteria: characteristic lipids[a].

Type of Bacteria	Characteristic Lipids
Archaebacteria	Various acyclic isoprenoid alkanes.
Methylotrophs	4-Methylsteroids, hopanoids.
Cyanobacteria (blue-green algae)	Hopanoids, mid-chain branched alkanes.
Sulfate reducers	Branched alkenoic acids.
Nitrosomonas	Hopanoids.
Nitrobacter	None known.
Metal reducers	None known.
Metal oxidizers	None known.

[a]These examples are purely illustrative; lipid composition often differs at the species level (e.g., for methanogens (10).

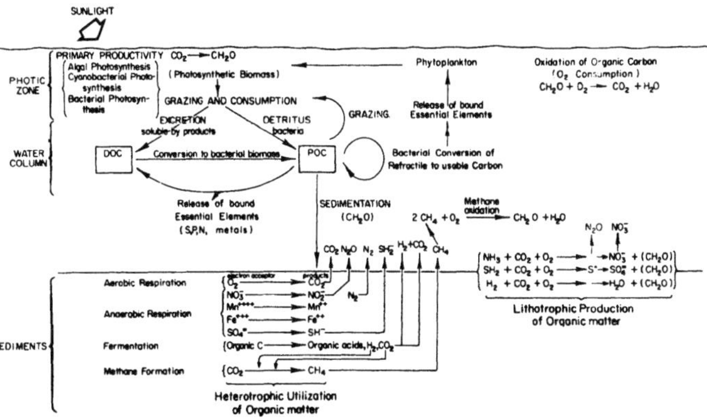

FIG. 1 - Known and postulated roles of microbes in the sea (Nealson).
POC - particulate organic carbon; DOC - dissolved organic carbon. Eh ranges: aerobic respiration - +500 to +800 mV; nitrate reduction - +300 to -500 mV; metal reduction - +100 to +400 mV; sulfate reduction - -100 to -400 mV; methanogenesis - -800 mV.

organic molecules can be modified by diagenetic processes. At an early stage, diagenesis can be biologically mediated, but the extent of the role of microbiota in this respect is unclear. Despite such limitations, the structural and stereochemical features of lipids in sediments can be used to infer their origin and history. For example, 17βH, 21βH-pentakishomohopane is probably derived from a functional precursor, tetrahydroxybacteriohopane, with an identical carbon skeleton.

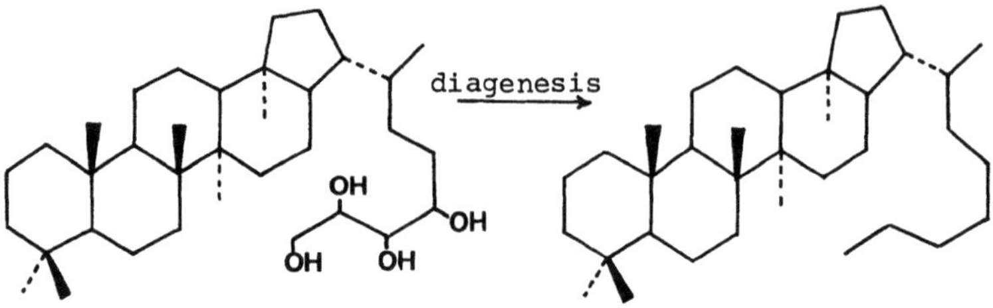

Tetrahydroxybacteriohopane 17ßH,21ßH-pentakishomohopane

The sources of sedimentary lipids are, however, often unclear: even so-called "marker" compounds may be derived from several sources. Bacterial reworking of lipids is important and is reflected in the composition of sedimentary lipids. It is not always clear whether direct microbial input rather than reworking of deposited material is the source of particular compounds, and initial benthic bacterial lipid signatures may well be reworked further down in the microbial hierarchy of sediments. Hence, we are frequently unsure of the extent to which the composition of sedimentary lipids represents original sediments inputs rather than the products of microbial reworking.

Non-lipids are also undoubtedly important microbial indicators but their full significance is unclear. Carotenoids have beer used with some success as input markers (e.g., (3)). Porphyrins are generally believed to be relics of photosynthetic activity, but they cannot, as yet, be used to define specific input sources. Other classes of compounds, such as amino acids, phospholipids, sugars and polysaccharides, and polymers such as cutins, lignins, and humic acids are geochemically significant but have not been extensively studied. They comprise a significant part of the organic matter in benthic sediments and of the kerogens in ancient sediments.

The lipid composition of prokaryotes and eukaryotes has been used to infer the evolution of biosynthetic pathways and of cell membranes (20). Organic geochemical analyses of sediments

from relevant geological eras and environments can provide evidence in support of these inferences. Methanogens, which are "primitive" archaebacteria are apparently unable to cyclize squalene; squalene therefore represents their most "advanced" terpenoid skeletal type. In the evolution of membranes, Ourisson et al. (20) postulate that "rigidifiers" follow the general evolutionary sequence:
biphytanyl ethers (prokaryotes) ⟶ carotenoids (prokaryotes) ⟶ hopanoids (prokaryotes) ⟶ steroids (eukaryotes).

Palaeontological evidence testifies to the evolution of plants and animals, but evidence of microbial evolution is scant. It is not known whether microbial populations have been as diverse in the past as they are at present and whether the effects of the benthic microbiota have been essentially constant. The search for morphological remains of microorganisms has been most thorough in Precambrian sediments. Does organic geochemistry provide the best means of evaluating microbial activity during the Phanerozoic?

Specific features of lipid distributions are remarkably similar in many sediments. For instance, the carbon number ranges in hopanoid distributions are similar in sediments from most ages and locations, whereas methanogenic lipids appear to be less widely distributed. The reason for these differences is not known.

Many sedimentary lipids (geolipids), particularly polyfunctional molecules, remain uncharacterized; advances in analytical techniques (particularly HPLC) and the synthesis of appropriate standards will lead to a fuller understanding of the organic geochemistry of sediments and hence of the lipid-biota relationships.

INTERACTIONS BETWEEN SEDIMENTS AND THE WATER COLUMN

The input of biologically derived material from the water column to sediments is of paramount importance for the organic geochemistry of sediments. Minute particles, such as planktonic remains, sink only very slowly through the water column; there is, therefore, ample time for the degradation and recycling of these particles. Large particulates (e.g., fecal pellets) sink rapidly through the water column. Transportation of organic matter to the sediment by fecal pellets should therefore be significant; this inference is supported by the presence of large numbers of fecal pellets in sediment traps and bottom sediments. Fecal pellets from the copepod Calanus helgolandicus contain small quantities of lipids whose composition qualitatively resembles that of the copepod's algal diet.

Terrestrial sources are also important contributors of organic matter to sediments. Evidence for this comes in part from the identification of visible plant fragments and cellular debris (e.g., pollen and spores), and from the presence of specific terrigenous marker lipids (e.g., cyclic diterpenoids).

The fraction of the terrestrial input of organic matter which reaches the sea floor must depend on many factors. The sedimentation rate of this material depends on its physical properties and on processes such as up- and downwelling in the water column. Other important influences, particularly in warm waters, are microbial activities and the oxicity/anoxicity of the water column. Questions which need to be asked include: 1) "How selective is the sedimentation process in the water column - as reflected in the organic debris reaching the bottom sediments? and 2) "Do fecal pellets comprise a major proportion of the input of organic matter from the euphotic zone?

RELATIONSHIPS OF METALS TO ORGANIC COMPOUNDS AND MICROORGANISMS

Although the metal cations dissolved in natural waters have been studied extensively, especially in seawater (e.g., (30)), little attention has been paid to the interaction of metals with organic substances and microorganisms.

The humic materials dissolved in natural waters are of interest not only because of their involvement in marine food chains and organic geochemical cycles (19) but also because of their ability to form complexes with metals (25,27). The difficulties in isolating and characterizing the humic materials, and the lack of techniques for examining the interactions between metals and polydentate ligands under natural conditions have yet to be overcome. The details of the structure of humic acids are unknown, but humic-metal binding is believed to be due to chelation by hydroxyl and carboxyl groups. Consideration of the stability constants of metal complexes of humic acids isolated from natural waters has led Mantoura et al. (16) to suggest that competitive complexation is important for the speciation of metals in natural waters. The proportion of humic-bound trace metals falls as salinity increases, possibly owing to competition from Ca^{+2} and Mg^{+2}, although the stability constants of the complexes of these ions with humic acids are relatively small.

Many other organo-metallic complexes are known to exist in natural aquatic systems. Porphyrin pigments such as heme, chlorophyll, and vitamin B are common in living organisms, and the existence of metalloporphyrins in ancient sediments is extensively documented (e.g., (22)). The siderochrome or ferrichrome group of compounds is an example of non-porphyrin organometallic complexation (28); these compounds are released by iron-requiring bacteria. They are widely distributed in organisms and are believed to be cofactors in microbial iron metabolism.

The study of the role of microorganisms in the deposition of metals began only recently. Experiments have demonstrated that certain bacteria can oxidize manganese (II) to manganese (IV) and precipitate MnO_2 (18). The widespread occurrence of these organisms and their ability to tolerate environmental extremes suggests that they may be of global significance in manganese precipitation at the benthic boundary layer. Manganese oxidation appears to be favored by low oxygen levels (microaerophilic), whereas growth is favored by high oxygen levels. Nutrients appear to exert complex and diverse effects on manganese oxidation. The biochemical mechanisms of manganese precipitation is still uncertain. As yet, no manganese oxidase enzyme has been isolated. The question: "Are manganese nodules formed by microbial processes?" is still unanswered.

The precipitation of metal sulfides by sulfate-reducing bacteria has been demonstrated (8). Many sedimentary ore deposits consist of metal sulfides, and microbial processes may have been important in their deposition. Anoxic sediments, such as the lower layers of microbial mats, which are rich in H_2S as a result of the activities of sulfate-reducing bacteria, may act as "traps" for metals dissolved in water (23). When water permeates the H_2S-rich region, insoluble metal sulfides are precipitated. The recent discovery of three new types of iron-oxidizing (Fe^{2+} to Fe^{3+}) bacteria lends support to the growing belief that microorganisms are important mediators of metal deposition. The newly discovered organisms are branched filamentous bacteria (200 μm in length), which possess localized deposits of iron (III) along the cell wall; they form an oxidizing layer at the sediment-water interface which overlies anoxic sediment. Microbes may also be indirectly responsible for metal deposition (22). That metals play a role in lipid binding within sediments is illustrated by the different composition of lipids extracted from a sediment with and without prior treatment with the metal complexing agent EDTA.

TEST MODELS: EXPERIMENTAL SYSTEMS

Several approaches can be made to the experimental study of the biogeochemical processes occurring in benthic sediments.

Open In Situ Incubations

This method has been used to follow the short-term diagenetic fate of a variety of lipids (7,13). Ths labelled substrate is introduced into a sediment core, which is then replaced in its environment for the duration of the experiment. Important diagenetic conversions demonstrated in this way include the oxidation of phytol to phytanic acid, and the hydrogenation of cholest-5-en-3β-ol to 5α- and 5β-cholestan-3β-ol.

"Cage"-type Experiments

In such experiments, selective exchange is allowed to take place between the incubated sample and its surroundings. For example, microbes may be incubated in a dialysis bag which is permeable to nutrients and metabolites, but which does not allow microbes to escape (24). "Big bag" (e.g., $100m^3$) experiments have been set up in open waters to study changes on a larger scale. The Alvin lunch-box (11) was an inadvertent cage-type experiment, since it was permeable to microbes but not to larger organisms. The disadvantage of this method is that the cage's permeability may be selective for particular substances and therefore may distort growing conditions.

In Situ "Box"-type Experiments

Incubations in such experiments are isolated, and no exchange with the environment takes place. This method has been used in studies of the metabolism of deep sea bacteria (11). It has the disadvantage that metabolites can accumulate and may affect the metabolic pathways of organisms.

Laboratory Incubations of Samples

These techniques share the advantages and disadvantages of in situ "box"-type incubations; they have the additional drawback that natural conditions such as temperature, pressure, and illumination and their cycles are difficult to reproduce. However, convenience is an asset, and laboratory incubation has been used successfully in conjunction with open in situ incubations in experiments involving the diagenesis of radiolabelled lipids (7,15), as well as in a number of microbial studies (11).

Culturing and Isolation of Organisms

Examples of such studies are numerous and diverse. They include studies of microbial manganese oxidation (18), fecal pellet production, deep sea barophiles (31), and "decay" experiments with specific contributing organisms. In addition, Vinogradski columns provide the opportunity to profile a layered system. The determination of the lipid composition of specific organisms is vital to the "biological marker" approach in organic geochemistry. However, recent work on phytoplankton (1) has shown that growth conditions can exert complex effects on the lipid composition of laboratory cultures. Caution must therefore be exercized when data for the lipids in sediments are compared with those for cultures.

Cell-free Systems

These systems involve breaking open cell walls or membranes, so that biochemical transformations can be observed free of constraints, such as membrane permeability, of cellular systems. Enzyme isolation can often be achieved from cell-free systems (e.g., cholesterol oxidase; (26)) and study of these enzymes can lead to a detailed knowledge at the molecular level of the mechanism of particular biochemical processes. However, in vivo work on whole cells by ^{13}C NMR has demonstrated that associations between enzymes in specific cellular sites are important, and thus cell-free systems may not reflect the natural behavior of cells.

Limitations of Labelling Studies

All feeding and culturing experiments perturb natural systems to a greater or lesser extent. The results of feeding experiments with radiolabelled compounds may be inaccurate because these substrates may be intracellular rather than extracellular in natural systems. When diagenetic processes are being studied, in which the lipids released by organisms can be acted on by other organisms, this is presumably not a great disadvantage. Experimental transformations may occur at different rates and may yield products different from those in unperturbed processes. However, if a product is predictable on the basis of the composition of a "natural" sample and if a transformation is biochemically reasonable, then it is usually safe to assert that the product is genuine. The question: "Is it a genuine incubation product?" is impossible to answer with absolute certainty. Novel products have been identified subsequently in natural, unperturbed samples.

It is of prime importance to determine whether or not an observed transformation is biological; biological processes can be recognized from their temperature optima (2). Control experiments are required to determine whether or not the chemical analytical procedures used in a particular set of experiments produce artefacts.

The most common method of labelling involves the use of radioisotopes. As a tracer ^{14}C is generally preferable to ^{3}H in biogeochemical studies, as ^{3}H can be readily lost by proton exchange processes. Isotope effects are also far greater for ^{3}H (and ^{2}H) relative to ^{1}H than for the corresponding carbon isotopes. However, labelling with ^{2}H or ^{3}H does allow the fate of individual substituent hydrogen atoms to be traced in biological molecular transformations.

The major disadvantage of radiolabelling is that, although the radioisotope is easy to detect, unambiguous identification of a radiolabelled product in a complex extract can be very difficult even with capillary radio-GC and radio HPLC. GC-MS cannot be used to detect a radioisotope unless it is present at undesirably high levels. The use of both stable- and radioisotopic labelling should overcome this drawback; ^{13}C and ^{2}H labelling has the additional advantage that NMR can be used to give structural information on isolated products.

The position of labelling is of fundamental importance in determining which transformation products can be observed. A label introduced by synthetic methods is often in a "vulnerable" position, i.e., it is subject to attack by organisms. For instance, the microbial conversion of cholest-5-en-3β-ol to 5-oxo-3, 5-seco-A-norcholestan-3-oic acid (29) is a well-known process which occurs with the loss of the carbon atom at position 4. Hence, for 4-^{14}C cholest-5-en-3β-ol, this keto-acid will not be a labelled product.

In any type of labelled incubation experiment, the "loading" aspect is critical. If natural levels of compounds are significantly exceeded, the metabolic pathways of the organisms may be changed and anomalous products will be generated.

SUMMARY

The nature of microbial activity in aquatic systems and bottom sediments is poorly understood. Some processes are mediated by bacteria, but their global significance is unclear. It is not known whether deep sea microbial processes are significant compared to those in surface waters and shallow waters, and it is not sure whether benthos is relatively inactive. Technical developments are needed to aid in the collection, culturing and study of deep sea bacteria without decompression. Little is known about the variations of microbial communities with depth within sediments. Questions such as: "Do hierarchies generally exist?"; "What is the extent of the interaction between microorganisms and their environment?"; "What limits the interactions?"; and "How variable are benthic environments and what factors control their biological activity within them?" are waiting to be answered.

Benthic microbiota may play a key role in the deposition of metals in aquatic environments, but little information is available, even for these processes today.

It has been demonstrated that abundant molecular evidence of past biological activity survives in aquatic sediments. The organic geochemical composition of sediments reflects the input to the sediment from the water column and the effects of the external environment, together with the contributions made and the changes brought about by the sedimentary biota, especially the microbial population. What is now needed is the ability to interpret this molecular evidence qualitatively and, perhaps, quantitatively, in terms of particular benthic microbial ecosystems. Such interpretations, set within the framework of particular types of deposit such as sedimentary iron ores, evaporites, phosphorites, and stratified sulfides bear directly on the goal of this conference. In particular, model ecosystems selected and controlled for the examination of specific microbial assemblages could provide appropriate parallels for the more extreme environments, past and present. The limited knowledge of microbial lipid biochemistry precludes the full understanding of the significance of the molecular evidence contained in sedimentary lipid "signatures." Questions such as: "Why are these lipid "signatures" so similar from many locations and geological eras?"; "What features of lipid composition are important indicators of past and present microbiological activity?"; "Are sedimentary lipid profiles largely controlled by the microbiota?"; "What are the major microbiological diagenetic pathways and how does abiological diagenesis affect the overall picture?" require further studies.

At present, many environments are relatively unexplored biogeochemically and much fundamental information remains to be uncovered. More data on sedimentary and microbial lipid composition are needed, together with information from interactive experiments such as incubations with isotopically labelled compounds.

Acknowledgements. We are grateful to K.H. Nealson for advice and access to unpublished work. We also thank I. Butler, F. Cameron, A. Dickson, D. Hamilton, G. Kenyon, J. de Leeuw, R.F.C. Mantoura, R. Morris, N. Owens, M.M. Quirk, M. Whitfield, and J.K. Volkman for helpful discussion. KE wishes to thank the SRC for a studentship. We thank the Natural Environment Research Council (GR3/2951) and the National Aeronautics and Space Administration (subcontract from NGL 05-003-003) for financial support. This paper is designated as a contribution to IGCP Project 157.

REFERENCES

(1) Ballantine, J.A.; Lavis, A.; and Morris, R.J. 1979. Sterols of the phytoplankton - effects of illumination and growth stage. Phytochem. 18: 1459-1466.

(2) Brock, T.D. 1978. The poisoned control in biogeochemical investigations. In Environmental Biogeochemistry and Geomicrobiology. III: Methods, Metals and Assessment, ed. W.E. Krumbein, Ch. 57, pp. 717-725. Ann Arbor Science.

(3) Cardoso, J.N.; Watts, C.D.; Maxwell, J.R.; Goodfellow, R.; Eglinton, G.; and Golubic, S. 1978. A biogeochemical study of the Abu Dhabi algal mats: a simplified ecosystem. Chem. Geol. 23: 273-291.

(4) Cohen, Y.; Paden, E.; and Shilo, M. 1975. Facultative anoxygenic photosynthesis in the Cyanobacterium Oscillatoria limnetica. J. Bact. 123: 855-861.

(5) Corliss, J.B.; Dymond, J.; Gordon, L.I.; Edmond, J.M.; Von Herzen, R.P.; Ballard, R.D.; Green, K,; Williams, D.; Bainbridge, A.; Crane, K.; and Van Andel, T.H. 1979. Submarine thermal springs on the Galapagos Rift. Science 203: 1073-1083.

(6) Didyk, B.M.; Simoneit, B.R.T.; Brassell, S.C.; and Eglinton, G. 1978. Organic geochemical indicators of paleoenvironmental conditions of sedimentation. Nature 272: 216-222.

(7) Gaskell, S.J., and Eglinton, G. 1975. Rapid hydrogenation of sterols in a contemporary lacustrine sediment. Nature 254: 209-211.

(8) Hallberg, R. 1978. Metal-organic interaction at the redoxcline. In Environmental Biogeochemistry and Geomicrobiology. III: Methods, Metals, and Assessment, ed. W.E. Krumbein, Ch. 76, pp. 947-953. Ann Arbor Science.

(9) Hodson, R.E.; Carlucci, A.F.; and Azam, F. 1979. Glucose transport in a low nutrient bacterium. Abstr. Am. Soc. Microbiol. N 59: 189.

(10) Holzer, G.; Oró, J.; and Tornabene, T.G. 1979. Gas chromatographic-mass spectrometric analysis of neutral lipids from methanogenic and thermoacidophilic bacteria. J. Chromatogr. 186: 873-877.

(11) Jannasch, H.W., and Wirsen, C.O. 1977. Microbial life in the deep sea. Sci. Am. 236: 42-52.

(12) Javor, B.J. 1979. Ecology, physiology and carbonate chemistry of blue-green algal mats, Laguna Guerrero Negro, Mexico. Ph.D. Thesis, Department of Biology and Graduate School of the University of Oregon.

(13) Javor, B.; Brassell, S.C.; and Eglinton, G. 1979. A laboratory/field method for radiolabelled incubation studies in algal-bacterial mats and other microbial ecosystems. Oceanologica Acta 2: 19-22.

(14) Krumbein, W.E.; Buchholz, H.; Franke, P.; Giani, D.; Giele, C.; and Wonneberger, K. 1979. O_2 and H_2S coexistence in stromatolites. Naturwissenschaften 66: 381-390.

(15) de Leeuw, J.W.; Simoneit, B.R.; Boon, J.J.; Rijpstra, W.I.C.; de Lange, F.; van der Laeden, J.C.W.; Correia, V.A.; Burlingame, A.L.; and Schenck, P.A. 1977. Phytol derived compounds in the geosphere. In Advances in Organic Geochemistry 1975, eds. R. Campos and J. Goni, pp. 61-79. Madrid: ENADIMSA.

(16) Mantoura, R.F.C.; Dickson, A.; and Riley, J.P. 1978. The complexation of metals with humic material in natural waters. Estuarine and Coastal Marine Science 6: 387-408.

(17) Margulis, L.; Barghoorn , E.S.; Ashendorf, D.; Banerjee, S.; Chase, D.; Francis, S.; Giovannoni, S.; and Stolz, J. 1980. The microbial community in the layered sediments at Laguna Figueroa, Baja California, Mexico: does it have Precambrian analogues? Precambrian Res. 11: 93-123.

(18) Nealson, K.H., and Ford, J. 1980. Surface enhancement of bacterial manganese oxidation: implications for aquatic environments. Geomicrobiol. J. 2: 21-37.

(19) Nissenbaum, A., and Kaplan, I.R. 1972. Chemical and isotopic evidence for the in situ origin of marine humic substances. Limnol. and Oceanog. 17: 570-582.

(20) Ourisson, G.; Albrecht, P.; and Rohmer, M. 1979. The hopanoids. Palaeochemistry and biochemistry of a group of natural products. Pure Appl. Chem. 51: 709-729.

(21) Owens, J.P.; Christofi, N.; and Stewart, W.D.P. 1979. Primary production and nitrogen cycling in an estuarine environment. In Cyclic Phenomena in Marine Plants and Animals, eds. E. Naylor and R.G. Hartnoll, pp. 249-258. Oxford: Pergamon.

(22) Quirke, J.M.E; Eglinton, G.; and Maxwell, J.R. 1979. Petroporphyrins I. Preliminary characterisation of the porphyrins of Gilsonite. J. Am. Chem. Soc. 101: 7693-7697.

(23) Renfro, A.R. 1974. Genesis of evaporite-associated metalliferous deposits - a sabkha process. Econ. Geol. 69: 33-45.

(24) Sakshaug, E., and Jensen, A. 1978. The use of cage cultures in studies of the biochemistry and ecology of marine phytoplankton. In Oceanogr. Mar. Biol. Ann. Rev., ed. H. Barnes, vol. 16, pp. 81-106. London: George Allen and Unwin.

(25) Siegal, A. 1971. Metal-organic interactions in the marine environment. In Organic Compounds in Aquatic Environments, eds. S.D. Faust and J.V. Hunter, pp. 265-295. New York: Marcel Dekker.

(26) Smith, A.G., and Brooks, C.J.W. 1976. Mini-review. Cholesterol oxidases: properties and applications. J. Steroid Biochem. 7: 705-713.

(27) Stumm, W., and Brauner, D.A. 1975. Chemical speciation. In Chemical Oceanography, 2nd ed., eds. J.P. Riley and G. Skirrow, vol. 1, pp. 173-234. London: Academic Press.

(28) Stumm, W., and Morgan, J.J. 1970. Aquatic Chemistry, pp. 288-291. New York: Wiley.

(29) Turfitt, G.E. 1948. The microbiological degradation of steroids 4. Fission of the steroid molecule. Biochem. J. 42: 376-383.

(30) Whitfield, M. 1975. Sea water as an electrolyte solution. In Chemical Oceanography, 2nd ed., eds. J.P. Riley and G. Skirrow, vol. 1., pp. 44-162. London: Academic Press.

(31) Yayanos, A.A.; Dietz, A.S.; and van Boxtel, R. 1979. Isolation of a deep-sea barophilic bacterium and some of its growth characteristics. Science 205: 808-810.

Mineral Deposits and the Evolution of the Biosphere, eds. H.D. Holland and M. Schidlowski, pp. 51-66. Dahlem Konferenzen, 1982.
Berlin, Heidelberg, New York: Springer-Verlag.

Microbiological Oxidation and Reduction of Iron

K. H. Nealson
Scripps Institution of Oceanography, La Jolla, CA 92093, USA

Abstract. Many bacteria oxidize iron and deposit iron oxides or hydroxides as insoluble cell-associated precipitates. Some produce intracellular magnetite. The acid tolerant oxidizers use the energy of oxidation for growth. The neutral pH iron oxidizers are generally heterotrophs. The morphology of heterotrophic iron oxidizers often resembles forms seen in recent and ancient iron deposits. Whether or not bacteria form extensive natural iron precipitates is not known, but they are almost always associated with such precipitates. The physiology of iron oxidizing bacteria is not well understood and this precludes extrapolation to pre-Phanerozoic sedimentary iron deposits.

INTRODUCTION

A causative role for bacteria in the deposition of iron oxides was suggested long ago (7), primarily on morphological grounds. A large number and variety of bacterial morphotypes are invariably associated with both ancient and modern iron oxide precipitates. This, coupled with "the almost universal presence in nature of organisms capable of precipitating iron from solution," led Harder to the conclusion that organisms were important in the formation of iron ore deposits. However, the role of bacteria in the oxidation and deposition of iron is difficult to prove even under controlled laboratory conditions. The problem is at least in part due to the nature of the process being studied. Ferrous iron (Fe^{++}), the geochemically mobile form, is readily autooxidized to ferric iron in the presence of even small amounts of molecular oxygen at pH values greater than 6.0, and ferric iron spontaneously precipitates as a constituent of one of a variety of oxides, hydrated oxides, or hydroxides. Thus, while the association of iron oxides with many different groups of bacteria is well documented, the fact that iron oxidizes spontaneously under the conditions in which these bacteria grow has left unanswered the question whether bacteria are necessary for, or even directly involved in, the formation of these precipitates. In order to assess the role of the biota, one must make accurate measurements of oxidation rates in the presence and absence of living microbes; "poisoned control" experiments must be performed. However, the behavior of iron is sensitive to the presence of microenvironments and to small changes in pH or Eh; iron is also very surface active, and arguments based on thermodynamic data are frequently misleading.

Iron is the fourth most abundant element in the earth's crust; its average crustal concentration is five percent. A good deal of this iron is present as a constituent of iron oxides. The most important sedimentary ores of iron are found in the Precambrian banded iron formations (BIF's), the origin of which is one of the topics of this symposium. In present day environments iron accumulates in bog iron ores, in ferromanganese nodules, and in Sabkha deposits; these accumulations are, however, small in comparison to the deposits in BIF's.

In spite of its considerable crustal abundance, most iron is unavailable to the biota because it cannot be taken up as a constituent of insoluble precipitates. Presumably, the microbiota (bacteria and fungi) which have biochemical mechanisms for obtaining iron play a central role in making iron available to plant and animal communities.

A review of the microbial interactions with iron is presented here, with emphasis on the following: a) mechanisms of iron uptake, b) mechanisms of iron oxidation, c) the physiology of iron oxidizing bacteria and their precipitates. The information that has been assembled may be useful in defining the role that organisms have played in weathering and in the deposition of pre-Phanerozoic iron formations.

CHEMISTRY

The complicated chemistry of iron oxidation will only be discussed briefly. At the earth's surface and in equilibrium with atmospheric oxygen, iron should be largely in the ferric state. However, the kinetics of iron oxidation are slow at pH values less than 6 (20). Above pH 4.5, the rate of Fe^{++} oxidation is given by the equation

$$-\frac{d\ (Fe^{++})}{dt} = k\ (Fe^{++})\ (OH^{-})^{2}\ P_{O_2}$$

where $k = 8\ (\pm\ 2.5)\ 10^{13}\ \text{mole}^{-2}\ \text{atm}^{-1}\ \text{min}^{-1}$.
Thus, a change of one pH unit produces a 100-fold change in the rate of Fe^{++} oxidation. When $P_{O_2} = 0.2$ atm and $Fe^{++} = 5$ ppm, the rate of oxidation of Fe^{++} is 8×10^{-6} ppm min^{-1} at a pH of 4.5, and 0.8 ppm min^{-1} at pH 7. It is therefore possible to study the oxidation kinetics of Fe^{++} relatively easily at low pH and hence to distinguish between the effects of chemical and of biological processes. On the basis of such rate data the microbiota have been implicated as causative agents in iron deposition in acidic environments such as bogs (4) and in acid mine drainage (18). At neutral pH, the situation is much less clear; spontaneous oxidation occurs so rapidly that the importance of microbial contributions to iron oxidation has not been determined. A wide variety of organic compounds such as hemin, EDTA, and dicarboxylic acids chelate ferric iron, and stabilize it in solution. Organic chelates, such as ferric ammonium citrate, are commonly used in the laboratory to supply iron in solution to aerobic organisms. In environments that are rich in organic compounds, such organic chelators probably alter the chemistry of iron significantly and complicate the interpretation of precipitation processes.

IRON AS AN ESSENTIAL ELEMENT

All life forms, with the exception of some lactic acid bacteria (14), require iron. Its major cellular function is the transfer of electrons, which were originally derived from growth substrates, between various cellular electron carriers and the harvesting of the energy from these electrons for biosynthesis or cell functions. Iron easily changes its valence state, and this process is well defined in inorganic systems. However, when complexed to organic ligands, the potential at which the valence change occurs can be dramatically different. Thus, iron is found as the central electron transfer metal in proteins between an electron potential of -400 and +300 mv (Table 1). Porphyrin derivatives such as heme (which are found bound to proteins in cytochromes), catalase, and peroxidase are common iron-containing ligands. The heme iron proteins are usually of higher potential than iron-sulfur and non-heme iron proteins. The iron-sulfur proteins are characteristically of low electron potential, and therefore allow electron transfer between highly electronegative couples. Ferredoxins, a class of iron sulfur proteins, have a wide range of cellular uses, including protection against the toxic properties of molecular oxygen.

IRON TRANSPORT IN BACTERIA

Because of the universal requirement for iron in cellular metabolism, most organisms require iron in the growth medium at concentrations of 0.4 to 4 micromolar. Since natural concentrations are usually about a factor of 10 below this level, microorganisms have developed specific methods for the uptake of iron. Siderochromes, or siderophores, are a class of compounds synthesized by bacteria and fungi that are excreted into the growth medium to facilitate iron uptake; probably all aerobic and facultatively anaerobic microbial cells produce or require siderochromes (14). Siderophore production is usually induced by growth on iron deficient media. Siderophore molecules are excreted into the growth medium and are taken up by cells only when Fe^{+++} is bound to them. They are usually quite specific for Fe^{+++}, and the complexes they form are quite stable (formation constants on the order of 10^{30}); when siderophore-Fe^{+++} complexes are taken up by organisms, iron is released by one of several mechanisms. The siderophores can be broadly grouped into two classes of compounds: the phenolic and the hydroxamate derivatives (Table 2).

Table 1 - Some iron containing cellular constituents

Component	Type of iron	E_m(mV)[1]	function
Cytochrome a/a_3	heme iron	290	reduction of O_2
Cytochrome c	heme iron	220	electron transport
Cytochrome b	heme iron	50	electron transport
Catalase	heme iron	-207	H_2O_2 removal
Pseudomonas ferredoxin	iron-sulfur	-235	camphor hydroxylation
Clostridium ferredoxin	iron-sulfur	-300	pyruvate oxidation
Nostoc ferredoxin	iron-sulfur	-350	photosynthetic electron transport
Nitrogenase	iron-molybdenum		nitrogen fixation

[1]E_m indicates the midpoint of the electron potential of the redox couple of the Fe^{++}/Fe^{+++} pair in the particular association. The values vary for the components from different cells, and the numbers are meant only to illustrate the range of Em's that the iron pair can exhibit in association with proteins.

Virtually all bacteria will take up heme, and heme iron can be used to supply iron to bacterial cultures. The transport mechanisms for heme uptake are not understood. As shown in Table 2, other chelators can be used to stabilize iron and to make it available for the biota; whether such reactions are important in nature is not known. While the mechanisms considered here are of great importance to the organisms that must compete for iron in an iron-limited environment, they were probably insignificant for the deposition of iron oxides in the past. Their function is to sequester the small amounts of iron available in the oxidized aquatic environment and to transport it into cells, where it is released and used for biological functions. The effect of such processes on the formation of Precambrian iron oxide deposits can probably be dismissed.

Table 2 - Some biologically active iron binding and transport compounds

1. Nonspecific iron chelators citrate EDTA dicarboxylic acids	Such compounds are used in the laboratory to supply iron. Importance in nature is not known.
2. Specific iron chelators	
a) heme and heme derivatives	All bacteria take up heme, but the permease system is not understood.
b) siderophores	Many bacteria and fungi produce these compounds; genera listed below.
1. hydroxamates	
mycobactin	Mycobacteria
schizokinin	Arthrobacter
ferrichromes	Many fungal genera
rhodotorulic acid	Rhodotorula
aerobactin	Aerobacter
2. phenolates	
enterobactin	Salmonella
itoic acid	Bacillus

1. See references (9,14) for details.

MECHANISMS OF BACTERIAL IRON OXIDATION

Iron oxidation may be either direct or indirect: direct oxidation consists of enzymatic or cell-associated processes; indirect oxidation consists of processes through which environmental conditions are altered and where these alterations lead to iron oxidation. Since the oxidation of iron is so sensitive to changes in Eh and pH, it would be surprising if indirect mechanisms did not play a role. Heterotrophic bacteria and fungi participate by raising the pH, either via amino acid or organic acid utilization (19):

1) amino acid $\longrightarrow$ keto acid + NH_4^+ + OH^-

2) salt of an organic acid $\longrightarrow$ CO_2 + H_2O + OH^-

Other indirect effects can be exerted by the cyanobacteria and algae via oxygen production and CO_2 consumption; oxygen production raises the Eh, while CO_2 consumption can raise the pH by its effect on the carbonate buffer system.

Yet another indirect mechanism, and one that can easily be confused with a direct mechanism, is the use of organic iron chelates by heterotrophic bacteria. Many bacteria, including bacilli and pseudomonads (11), use the organic carbon of such ligands as a source of energy; the iron is therefore freed and is easily oxidized.

Direct oxidation of iron by bacteria can be divided into two categories: enzymatic oxidation and oxidation via cellular products that have not been proven to be enzymes. Bacteria in the first group are acid tolerant forms, some of which can use the energy of oxidation for growth; those in the latter group are characteristically associated with natural iron precipitates.

The only bacterium unequivocally proven to be a strict iron autotroph is the acid tolerant iron oxidizing Thiobacillus ferrooxidans. This organism, which grows well between pH 2.5 and 3.5, can exist with ferrous iron as its only source of energy and CO_2 as its sole source of carbon. Ribulosebisphosphate carboxylase is present, and carbon is presumably fixed via the Calvin-Benson cycle. Other bacteria suspected but not proven to operate in this manner are shown in Table 3. Thiobacillus is usually restricted to acidic environments; in these environments other thiobacilli oxidize sulfur and produce sulfuric acid which stabilizes ferrous iron in solution. In fact, several of the acid tolerant iron oxidizers can oxidize sulfur. It has been possible to establish the energetics for T. ferrooxidans and to prove that iron oxidation is a metabolic pathway of importance (13). T. ferrooxidans grows at the expense of iron oxidation, producing soluble ferric iron that precipitates when it reaches an area of high pH, or which may form $Fe(OH)_3$ by reaction with water. Large quantities of precipitated iron hydroxides or oxides are commonly found downstream from acid mines where the pH rises. Thiobacilli are probably not primitive organisms; they are obligate aerobes adapted to extreme conditions. They probably evolved after the full oxidation of the atmosphere and thus may not have been involved in the formation of pre-Phanerozoic iron ore deposits.

Table 3 - Some iron bacteria and their habitats.

		pH of growth
I. Acid tolerant forms		
	Thiobacillus	2-4
	Sulfolobus	2-4
II. Neutral pH forms		
Sheathed bacteria	Sphaerotilus-Leptothrix	6-8
	Clonothrix	6-8
Budding bacteria	Hyphomicrobium	6-8
	Pedomicrobium	6-8
	Gallionella	6-8
	Mycoplasma	6-8
	Metallogenium[1]	6-8

[1]Status of genus uncertain (see later discussion); one acid tolerant form reported (22).

The bacteria which oxidize at neutral pH are perhaps the most relevant group of organisms, but they are also the most controversial group. It has not been possible to demonstrate unequivocally that their growth is autotrophic or even that oxidation is enzymatic, but there are many organisms that can catalyze iron oxidation. In the presence of these organisms the rate of oxidation is significantly faster than in their absence. Some of these organisms are listed in Table 3; it is curious that most of the bacteria are structurally complex; sheathed, budding, and stellate forms predominate. Laboratory experiments - in which iron oxidation is enhanced by their presence - and field observations both suggest that these bacteria could have been involved in the formation of iron precipitates (7). Wolfe (10) has suggested that the neutral pH forms are "gradient organisms," capable of establishing themselves at oxygen levels where they can successfully compete with autooxidation processes. This hypothesis is consistent with the observation that members of the group live under conditions of low oxygen tension, i.e., that they are microaerophilic (15).

The sheathed bacteria, commonly known as the Sphaerotilus-Leptothrix group, are small, gram negative rods; they commonly form filaments surrounded by sheaths that can be heavily encrusted with iron and/or manganese. Some species oxidize either iron or manganese, while others oxidize both. While autotrophic growth on iron has been proposed, it has never been proven, and it seems likely that iron oxidation proceeds spontaneously while organic compounds are used as a source of energy. Many examples of sheathed bacteria in natural iron depostis are listed by Harder (7); recent ultrastructural studies show details of Leptothrix found in iron rich lake and cave deposits (3).

Budding bacteria form new growths (buds) as extensions of the cell and thus assume a variety of morphological forms (8). In this diverse group, many genera, including Gallionella, which has been known since the early 1800s, have oxidizing members. G. ferruginea forms small bean shaped cells with long associated stalks of $Fe(OH)_3$, presumably the excretion product of growth on ferrous iron. It is microaerophilic, preferring 0.1 - 1 $\mu g\ O_2\ l^{-1}$ (24). It has been considered an autotroph, as it a) requires Fe^{++} for growth, b) grows with Fe^{++} and very little organic carbon, and c) fixes CO_2 into cell carbon (10). However, unequivocal evidence of autotrophy has not been presented, and since the work of Wolfe, little has been done with Gallionella to complete the understanding of its physiology and biochemistry.

Pedomicrobium bacteria is commonly found associated with iron deposits in soils, lakes, and shallow marine basins; it can oxidize iron and manganese, although some species are restricted to the oxidation of one or the other metal. External polysaccharides may chelate the iron (6); if so, the cells probably do not use metal oxidation as an energy yielding reaction.

Perhaps the most controversial and interesting of the budding bacteria are those of the group Metallogenium. Several species of this genus have been described, including M. invisum, M. personatum, and M. symbioticum (8), but only a few workers have claimed success in obtaining pure cultures of these bacteria in the laboratory. The extant species are associated primarily with manganese oxides, although in nature, up to 30% of the afflicted deposits may be iron (15). The bacteria are characterized by a complex "life cycle" defined primarily on the basis of precipitate morphologies. The cycle consists of several stages ranging from single cells to complex stellate budding forms to large amorphous aggregates (15). While some workers have claimed success in culturing these bacteria and have placed them in the group Mycoplasmetales, others have not been able to reproduce the results and have challenged the validity of the genus (16). On the basis of field studies, many workers have implicated Metallogenium in the formation of iron and manganese deposits; the establishment of the genus and studies of the physiology and biochemistry of its members are therefore very important.

The interest in Metallogenium stems not only from its abundance in present day deposits, but in its similarity to fossil organisms as well. In studies of the Lower Gunflint Iron Formation, Barghoorn and Tyler (2) reported abundant microfossils remarkably similar to modern Metallogenium; these were called Eoastrion simplex. Similar microfossils have since been identified in many other ancient iron deposits; this supports the hypothesis that such organisms were important in the formation of the deposits. It has now been suggested on the basis of their morphological similarity that Eoastrion and Metallogenium are related (1).

What, if any, are the advantages of iron oxidation to bacteria? Bacteria may derive energy from iron oxidation, which is a distinct advantage in nutrient limited environments; this ability is known to be a property of the acid tolerant forms, but of no others. Iron oxidation may have developed as a mechanism of oxygen detoxification (removal). The presence of a simple cell-associated O_2 removal system, such as a sheath or other extracellular component, might have been of great advantage before the evolution of enzymatic mechanisms to protect cells from oxygen. The process could thus be viewed as an extracellular oxygen sink of no particular energetic advantage. However, the deposition of insoluble oxides around bacteria might be a significant disadvantage, and it is not surprising that the organisms commonly associated with iron oxidation are sheathed and budding forms that are able to escape from their precipitates. It should be noted that protection from oxygen toxicity is a postulated function of iron in several protein associations, including hemoglobin, iron-sulfur proteins, peroxidase, catalase, and some superoxide dismutases; the hypothesis that primitive extracellular oxygen protection mechanisms could have arisen via iron chelation is therefore tempting. Iron oxidation may, of course, offer many other advantages to cells, among them the stabilization of cell holdfasts and the formation of protective cell coverings (3).

IRON OXIDIZING BACTERIA: THEIR ROLE IN THE FORMATION OF BIF'S

Bacteria that catalyze iron oxidation indirectly by altering environmental conditions may have been important in the formation of ancient iron deposits; but this notion is hard to assess without accurate information regarding environmental conditions at the time of iron ore deposition. As the requisite paleoenvironmental data become available, it may be possible to assign plausible roles to bacteria in the deposition of iron. There is little doubt that abundant microbiota were living at the time of formation of the BIF's, but whether or not these organisms played causative roles in iron formation is not certain. It is easy to imagine that either of the first two functions discussed above for iron oxidation could have been advantageous and selected for once oxygen began to accumulate in the atmosphere. It is not obvious that there would have been a driving force for the development of the bacterial systems for iron deposition without the presence of free oxygen.

Nevertheless, the best and perhaps the only methods for understanding the origin of ancient iron deposits is by extrapolating our knowledge of present day microbes to their fossil analogues. The major problem with this approach is that there is a paucity of data concerning extant organisms and little basis for accurate extrapolation over long periods of time. Indeed, it is difficult to establish that the observed microfossils are bacterial remains at all, let alone iron bacteria. Laboratory and field studies are surely consistent with the hypothesis that bacteria could have played a role in the deposition of ancient iron deposits, but presently available evidence does not constitute proof. Continued study of the physiology and morphology of extant organisms and further examination of iron deposits, past and present, may clarify the matter. It might be asked whether genetic implications might be drawn from the mineralogy of the iron ore deposits. Are there, for instance, minerals that are produced exclusively by inorganic or biological means? The information is scanty, but some recently discovered magnetotactic bacteria are now known to deposit magnetite inside their cells. Magnetite was previously regarded as a mineral that could be formed only inorganically. All other extant iron depositing bacteria initially deposit iron hydroxides, although the mineralogy of these hydroxides is not well studied. Whether or not bacteria exist (or existed) that deposit magnetite or hematite, minerals commonly found in the banded iron formations, extracellularly or in bulk, is not known.

MICROBIOLOGICAL IRON REDUCTION

A final consideration in the deposition of iron oxides is the capacity of many bacteria to reduce oxidized iron under anaerobic conditions. This activity may be viewed as a biological weathering of iron deposits and is well documented in modern sediments, where iron is mobilized as a result of microbial activity in lake sediments (12). Bacterial iron reduction is poorly understood, although it is well documented that many bacteria can catalyze the reduction of ferric to ferrous iron (12,15). Whether the mechanisms are direct of indirect is not known, and further study of this group of organisms is surely warranted. This is especially true when one considers that one of the enigmatic features of the BIF's is their regular banding. It is possible that this banding is the result of alternating periods of iron dissolution or weathering as a result of bacterial activity. If disposition was a continuous process and weathering a cyclic one, iron poor bands would have been produced during dilution, leading to the formation of banded deposits. Without data to support one mechanism more than the other, the hypothesis that iron reducers are responsible for the banding patterns seen in the ancient deposits should not be abandoned.

REFERENCES

1) Barghoorn, E.S. 1977. Eoastrion and the Metallogenium problem. In Chemical Evolution of the Early Precambrian. New York: Academic Press.

2) Barghoorn, E.S., and Tyler, S.A. 1965. Microorganisms from the gunflint chert. Science 147: 563-577.

3) Caldwell, D., and Caldwell, S. 1980. Fine structure of in situ microbial iron deposits. Geomicrobial. J. 2: 39-53.

4) Crerar, D.A.; Knox, G.W., and Mans, J.L. 1979. Biogeochemistry of bog iron in the New Jersey pine barrens. Chem. Geol. 24: 111-135.

5) Dondero, N.C. 1975. The Sphaerotilus-Leptothrix group. Ann. Rev. Microbiol. 29: 407-428.

6) Ghiorse, W.C., and Hirsch, P. 1979. An ultrastructural study of iron and manganese deposition associated with extracellular polymers of Pedomicrobium-like budding bacteria. Arch. Microbiol. 123: 213-226.

7) Harder, E.C. 1919. Iron-depositing bacteria and their geologic relations. U.S. Geological Survey, Professional Paper No. 113. Washington, D.C.: Government Printing Office.

8) Hirsch, P. 1974. Budding bacteria. Ann. Rev. Microbiol. 28: 391-444.

9) Hutner, S.H. 1972. Inorganic nutrition. Ann. Rev. Micro-Biol. 26: 313-346.

10) Kucera, S. and Wolfe, R.S. 1957. A selective enrichment for Gallionella ferruginea. J. Bact. 74: 347-350.

11) Kullman, K.H., and Schweisfurth, R. 1978. Iron-oxidizing rod-shaped bacteria II. Quantitative study of metabolism and iron oxidation using iron (II) oxalate. Allg. Mikrobiol. 18: 321-327.

12) Kuznetsov, S.I. 1970. The microflora of lakes and its geochemical activity. Austin, TX: University of Texas Press.

13) Lundgren, D.G., and Dean, W. 1979. Biogeochemistry of iron. In Biogeochemical Cycling of Mineral Forming Elements, eds. P.A. Trudinger and D.J. Swaine, pp. 202-211. Amsterdam: Elsevier Press.

14) Neilands, J.B. 1973. Microbial iron transport compounds. In Inorganic Biochemistry, ed. G.L. Eichorn, pp. 176-202. Amsterdam: Elsevier Press.

15) Perfil'ev, B.V.; Gabe, D.R.; Gal'perina, A.M.; Rabinovich, V.A.; Sapotniskii, A.A.; Sherman, E.E.; and Troshanov, E.P. 1965. Applied Capillary Microscopy. New York: Consultants Bureau.

16) Schweisfurth, R.; Eleftheriadis, D.; Gunlach, H.; Jacobs, M.; and Jung, W. 1978. Microbiology of the precipitation of manganese. In Environmental Biogeochemistry and Geomicrobiology, ed. W. Krumbein, pp. 923-928. Ann Arbor, MI: Ann Arbor Science.

17) Silverman, M.P., and Ehrlich, H.L. 1964. Microbial formation and degradation of minerals. Adv. Appl. Micro. 6: 163-203.

18) Singer, P.C., and Stumm, W. 1970. Acid mine drainage: the rate determining step. Science 167: 1121-1123.

19) Starkey, R.L., and Halvorson, H.O. 1927. Studies on the transformation of iron in nature II. Concerning the importance of microorganisms in the solution and precipitation of iron. Soil Sci. 24: 381-402.

20) Stumm, W., and Morgan, J.J. 1970. Aquatic Chemistry. New York: Wiley Interscience.

21) vanVeen, W.L.; Mulder, E.G.; and Deinema, M.H. 1978. The Sphaerotilus-Leptothrix group of bacteria. Microbiol. Rev. 42: 329-356.

22) Walsh, F., and Mitchell, R. 1972. An acid-tolerant iron oxidizing Metallogenium. J. Gen. Microbiol. 72: 369-374.

23) Weinberg, E.D. 1978. Iron and infection. Microbiol. Rev. 42: 45-66.

24) Wolfe, R.S. 1963. Iron and manganese bacteria. In Principles and Applications in Aquatic Microbiology, pp. 82-97. New York: John Wiley.

25) Yoch, D.C., and Carithers, R.P. 1979. Bacterial iron-sulfur proteins. Microbiol. Rev. 43: 422-442.

Mineral Deposits and the Evolution of the Biosphere, eds. H.D. Holland and M. Schidlowski, pp. 67-82. Dahlem Konferenzen, 1982.
Berlin, Heidelberg, New York: Springer-Verlag.

The Pre-Phanerozoic Fossil Record

S. M. Awramik
Dept. of Geological Sciences, University of California
Santa Barbara, CA 93106, USA

Abstract. The pre-Phanerozoic fossil record is dominated by stromatolites and microfossils. The most ancient traces of life that are now confidently recognized are the stromatolites and filamentous microfossils from the ~3500 Ma-old Warrawoona rocks in Western Australia. Younger Archean (up to 2500 Ma-old) stromatolites, though rare, indicate that the benthic microbial ecosystem was well established. It is not known whether the Archean microbes that built the stromatolites or may have inhabited the water column were oxygen-releasing photoautotrophs. The fossil record becomes conspicuous during the Proterozoic (2500 to 570 Ma ago). Stromatolites, some with well preserved cyanobacterial-like microfossils are plentiful, and the variety of microfossils found in rocks from different sedimentary environments suggest a considerable diversity of microbial ecosystems. Phytoplankton represented by acritarchs apparently underwent a major radiation ~1400 Ma ago. Microfossils that have been interpreted as eukaryotes have been reported from a few Proterozoic sequences and are the subject of great debate. Centimeter-sized algal (eukaryote) remains have been found in rocks ~1300 Ma-old.

INTRODUCTION

Pre-Phanerozoic fossils can be grouped into three broad categories: 1) stromatolites, 2) algal megafossils, and 3) microfossils. Stromatolites are the most conspicuous fossils encountered by field geologists working in pre-Phanerozoic terranes. They are one of the most ancient traces of life known on Earth and have been found in 3500 Ma-old rocks (27,43).

Stromatolites existed throughout the remainder of the pre-Phanerozoic, through the Phanerozoic and are still forming today. By analogy to modern structures, fossil stromatolites are commonly viewed as biosedimentary structures produced by the sediment trapping, binding, and precipitating activities of cyanobacteria. Other bacteria can, however, also participate in stromatolite formation today and may have built the earliest stromatolites.

In the geological record, this type of microbial activity is commonly recorded in laminated domal and columnar structures in limestones and dolomites. Some of these structures are several meters in diameter. The same processes can also lead to less spectacular but more abundant planar and non-laminated constructions (38). Limestone and dolomite are the dominant hosts of stromatolites; however, pre-Phanerozoic stromatolites are also found in cherts (8), phosphorites (7), and iron-formations (45).

Macroscopic algae can be preserved as carbonaceous films several millimeters in diameter and are known from rocks as old as ~1300 Ma (44). Such algal remains are the most reliable evidence for the presence of eukaryotes. However, large algal fossils are rare in the pre-Phanerozoic record.

Microfossils from the pre-Phanerozoic fall into one of two categories: stromatolitic and non-stromatolitic. Both types are commonly preserved as organic-walled microstructures of varying size, shape, and organization. At present, the most useful microfossils for understanding ancient microbial ecosystems and for gaining insight into ancient biological processes are those found in silicified stromatolites. Their preservation may be remarkably good; much of the apparent original morphology of microbes is often preserved. Most microfossils found in stromatolites have modern morphological analogs among the cyanobacteria and, in a few cases, among the bacteria. Well preserved microfossils in silcified stromatolites have been found in about 100 localities. The majority of these stromatolitic microfossils is found in strata less than 1600 Ma-old (33). The oldest is in the ~3500 Ma-old Warrawoona Group of Western Australia, where wavy to planar, laminated probably stromatolitic cherts contain well-preserved filamentous microfossils (6).

Non-stromatolitic microbiotas, that include both benthonic and planktonic microfossils, are dominated by "acritarchs." By definition, these are organic-walled spheroids of uncertain taxonomic position (14,15). Non-stromatolitic microfossils are found primarily in shale, siltstone, arkose, and sandstone (41), as well as in chert (3) and diamictite (24). Other non-stromatolitic microfossils include pluricellular aggregates and filaments comparable to cyanobacteria, and radiating filamentous bundles which resemble bacteria in morphology and in organization.

The literature dealing with acritarchs and related microfossils is vast (41). Hundreds of formations, most 1400 Ma-old and younger, are reported to contain such microfossils. However, many of these putative microfossils may be micropseudo-fossils produced during laboratory preparation procedures and/or post-depositional microbial contaminants introduced by endolithic activity though cracks at some unspecified time after deposition or by means of simple contamination on the outcrop, at the collecting site, or enroute to or in the laboratory (12).

THE ARCHEAN

The Archean Eon (see Fig. 1) does not abound in fossils, but it does contain the earliest traces of life. The oldest well-studied Archean terrane containing metasediments is the ~3800 Ma-old Isua area of West Greenland (28). No compelling evidence for life is found in the metasedimentary rocks at Isua; however, these medium grade metamorphic rocks contain some information that is suggestive of biological activity at the time of their deposition. $\delta^{13}C_{org}$ values are consistent with photoautotrophy (31), if a poorly understood correction is made for

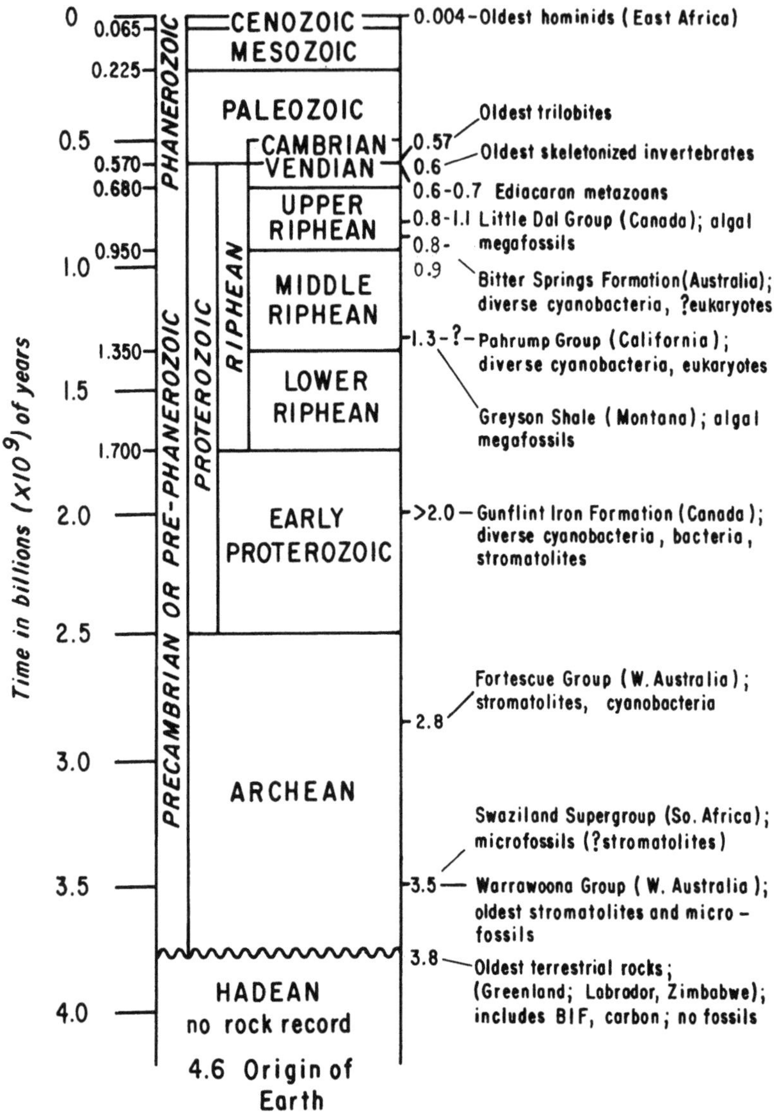
Time in billions (X10^9) of years
0
0.065
0.225
0.5
0.570
0.680
0.950
1.0
1.350
1.5
1.700
2.0
2.5
3.0
3.5
4.0
PHANEROZOIC
PRECAMBRIAN OR PRE-PHANEROZOIC
PROTEROZOIC
RIPHEAN
CENOZOIC
MESOZOIC
PALEOZOIC
CAMBRIAN
VENDIAN
UPPER RIPHEAN
MIDDLE RIPHEAN
LOWER RIPHEAN
EARLY PROTEROZOIC
ARCHEAN
HADEAN
no rock record
4.6 Origin of Earth
0.004 – Oldest hominids (East Africa)
0.57 – Oldest trilobites
0.6 – Oldest skeletonized invertebrates
0.6-0.7 Ediacaran metazoans
0.8-1.1 Little Dal Group (Canada); algal megafossils
0.8-0.9 Bitter Springs Formation (Australia); diverse cyanobacteria, ?eukaryotes
1.3-?- Pahrump Group (California); diverse cyanobacteria, eukaryotes
Greyson Shale (Montana); algal megafossils
>2.0 – Gunflint Iron Formation (Canada); diverse cyanobacteria, bacteria, stromatolites
2.8 Fortescue Group (W. Australia); stromatolites, cyanobacteria
Swaziland Supergroup (So. Africa); microfossils (?stromatolites)
3.5 – Warrawoona Group (W. Australia); oldest stromatolites and micro-fossils
3.8 – Oldest terrestrial rocks; (Greenland; Labrador, Zimbabwe); includes BIF, carbon; no fossils

FIG. 1.

metamorphic alteration of the $\delta^{13}C$ values and if it is assumed that light was the energy source for C-fixation reactions (J.M. Hayes, written communication). Banded iron formation (BIF) at Isua has been interpreted to indicate the presence of some free O_2; oxygen-releasing photosynthesis is the most likely source of free O_2 (10,11). However, no stromatolites or convincing microfossils are known from the Isua sequence (9).

The oldest unequivocal evidence of life comes from cherts of the ~3500 Ma-old Warrawoona Group of Western Australia. There, stromatolites (26,43) and filamentous microfossils (6) are known from cherts. Kerogenous spheroids have been found (16), but their interpretation from these and from younger rocks is controversial (6,36). The affinities of the filamentous microfossils are somewhat obscure; there are no diagnostic features characteristic of a single prokaryotic group (6).

Organic spheroids, which may be biogenic (36), are also known from the ~3500 Ma-old Onverwacht (29) and from the somewhat younger Fig Tree (22) Groups of South Africa and from the >2500 Ma-old Hamersley Group of Western Australia (26). Probable filamentous microfossils that are morphologically similar to specific types of modern cyanobacteria are known from the ~2800 Ma-old Fortescue Group of Western Australia (36) (Table 1). One must keep in mind the need to demonstrate the biogeneity of such ancient as well as younger microfossils.

The stromatolite record for the Archean is somewhat more substantial than the microfossil record. However, aside from the Warrawoona Group, only 8 other localities with convincing stromatolites are known (Table 1).

Thus, the 1300 million years of Archean time have yielded only a meager fossil record. Stromatolites are rare and microfossils are even rarer. The Warrawoona fossils do, however, indicate that benthic microbial communities existed and were diverse in composition and perhaps in physiology some 3500 m.y. ago.

TABLE 1 - Archean Fossils

Age (Ma)	Rock Unit	Location	Stromatolites	Micro-fossils
>2500	Steeprock Group	Ont., Canada	X	
>2500?	Turee Creek Group	W. Australia	X	
>2500	Hamersley Group	W. Australia	X	
~2640	Ventersdorp Group	South Africa	X	
~2640	Bulawayan Group	Zimbabwe	X	
~2650	Yellowknife Group	NWT, Canada	X	
~2700	Woman Lake Marble	Ont., Canada	?	
~2800	Fortescue Group	W. Australia	X	X
~3000	Insuzi Group	South Africa	X	
~3400	Fig Tree Group	South Africa		?
~3500	Onverwacht Group	South Africa	?	X
~3500	Warrawoona Group	W. Australia	X	X

data from (36) and (42)

THE PROTEROZOIC

The Proterozoic Eon spans nearly 2000 million years from 2500 to 570 Ma ago (20) (Fig. 1). The microfossil record in this eon is rich and diverse in both stromatolitic and nonstromatolitic sedimentary rocks. Stromatolites become a noticeable part of shallow-water marine carbonate deposits early in the Proterozoic. Columnar forms achieve maximum diversity and abundance during the Late Riphean (about 1000 Ma ago); their diversity decreases as the Cambrian is approached (1). About 1400 Ma ago the size of microfossils increased dramatically. This increase probably signalled the diversification of the eukaryotes (34).

PROTEROZOIC STROMATOLITES

The Proterozoic witnessed the acme of stromatolite development. The stabilization of cratons in the late Archean and early Proterozoic produced widespread shallow-water environments that led to a major benthic microbial radiation. By 2000 Ma ago, stromatolites were abundant, geographically widespread and morphologically diverse (domes; columns including unbranched, branched, and conical; nodular types; bioherms; oncolites; and planar or stratiform types).

The morphology of stromatolites is notoriously variable. However, in single outcrops, morphological themes representing narrower ranges of variability are commonly exhibited. In the late 1800s an effort was made to classify morphologically distinct stromatolites. In the 1960s their paleontological classification reached a high degree of sophistication, and it was demonstrated by Soviet scientists that certain stromatolite morphologies (defined as taxa assemblages of stromatolites and microstructures) had restricted time ranges in the Proterozoic, and therefore were useful as index fossils (25,37). Columnar, columnar-layered and conical stromatolites were found by many researchers to be best suited for these analyses. These findings were later confirmed with a fair degree of success in regions outside of the USSR (13). The fourfold subdivision of the Riphean in Fig. 1 is based on distinctive assemblages of stromatolites. The time-restricted stromatolites probably reflect the evolution of stromatolite-building microbes, but details of this microbial evolution and of the microbial control of stomatolite form are obscure (38). Unfortunately, except in a few cases, biostratigraphically useful columnar stromatolites do not contain microfossils. Hence, no convincing test of this evolutionary explanation is possible at present.

At the end of the Proterozoic there was a marked decrease in diversity of columnar stromatolite taxa (1). This decrease was probably due to the activities of detritus-feeding animals whose appearance is first recorded in the Vendian. Subtidal environments, the sites of the major development of biostratigraphically useful stromatolites (39), were also sites of early metazoan evolution and diversification.

PROTEROZOIC MICROFOSSILS

Stromatolitic Microbiotas

In an analysis of Proterozoic microbiotas described primarily from stratiform stromatolites, Schopf (33) found that spheroidal unicells, cylindrical tube-like sheaths, and cellular trichomic filaments show an increase in mean diameter and size accompanied by an increase in taxonomic diversity at about 1400 Ma ago. A greater diversity of filaments is found in younger Proterozoic stromatolites. Yet within this framework of change, one is struck by the high degree of morphological conservatism exhibited by many of the microfossils and by their recurrence in paleontologically recognizable microbial communities. For example, the Early Proterozoic microbiota from the Belcher Islands is surprisingly similar in composition and community structure to the 800-900 Ma-old microbiota from the Bitter Springs Formation, Australia (18). This is also true for some mid-Proterozoic microbiotas such as the Balbirini microbiota of Australia (30). Early in the study of pre-Phanerozoic microfossils, Schopf (32) noted the morphologic similarity between Proterozoic forms and modern cyanobacteria. The extraordinarily long morphological stagnation of many of these cyanobacteria-like microfossils appears to be at odds with the notion that stromatolite-building microorganisms evolved through time. However, on close inspection, one sees that the types of stromatolites used in the two studies are different. Almost all microfossiliferous stromatolites are of the planar or flat laminated variety which generally accumulated in periodically exposed environments but also ranged into very shallow, permanently submerged environments. Stratiform-stromatolite morphology is virtually invariant through geologic time and is therefore of little value in biostratigraphy. On the other hand, the columnar varieties of stromatolites which tend to be unmicrofossiliferous are biostratigraphically useful. Columnar stromatolites formed principally in subtidal settings (39). What little information we do have regarding microfossils in columnar stromatolites (2,35) suggests that changes in microbial community structure played a great role in determining shape at any one time. Probably the introduction of new species into microbial communities resulted in a change in community structure and, in turn, stromatolite shape.

MICROFOSSIL CHANGES THROUGH THE PROTEROZOIC

In order to evaluate evolutionary trends during the Proterozoic Eon, a good indication of the nature of the biota is required. The Gunflint and Belcher Islands microbiotas, both from Canada and both ~2000 Ma-old, supply the required data. Stromatolitic microfossils preserved in chert dominate both formations (3,8,18). In the Gunflint a cherty, non-stromatolitic facies with microfossils is also known (23). However, neither the Gunflint nor the Belcher contains microfossils from clastic facies. In stromatolites from both deposits, microfossils that are morphologically comparable with modern cyanobacteria and comprise ~99% of the microbiotas and more than 29 of the form taxa (= fossil species) are assignable to 4 of the 5 modern orders of cyanobacteria. Chroococcales- and Oscillatoriales-like morphs predominate in the stromatolites. In the Gunflint non-stromatolitic facies spheroidal unicells (up to 31 µm in diameter) and rosettes of radiating filaments (Eoastrion Barghoorn) dominate (3,23). Eoastrion is commonly compared morphologically and physiologically to the extant problematic, putative Mn-oxidizing bacterium Metallogenium (21); however, it can also be compared morphologically to the sheathed bacterium Leptothrix echinata Beger.

These microbiotas together with allied microbiotas in Western Australia (3) indicate that prokaryotic life, in particular cyanobacterial microbes, was well established, diverse, and geographically widespread during early Proterozoic time.

A major change in pre-Phanerozoic microbiotas took place ~1400 Ma ago. During a microbial transition which was first recognized by Timofeev with acritarchs (40) 1) new and large taxa appeared (32,33,38) while most of the earlier taxa on the generic level persisted; 2) localities yielding microfossils are more abundant; 3) clastic rock sequences (non-stromatolitic) become major sources of microfossils, in particular acritarchs; and 4) acritarchs, presumably phytoplankton, become widespread. The stromatolitic chroococcacean-oscillatoriacean (comparing the morphs to modern cyanobacteria) microbial community established earlier remained unchanged, but larger and more robust oscillatoriaceans played a greater role in mat construction. The appearance of acritarchs, which best records the microbial change 1400 m.y. ago, probably signalled the radiation of nucleated algae (34).

In late Vendian to Early Cambrian rocks, stromatolitic microfossils as well as acritarchs show some changes. Acritarchs become ornamented with spines. Stromatolitic filamentous sheaths and trichomes increase in diameter in the Vendian (33), but there are apparently no changes in the size of stromatolitic spheroidal microfossils. Presumably, selective pressures brought on by heterotrophic protists and the newly evolving metazoans favored larger, more robust filaments; many of the spheroids, which were already able to form rubbery, pluricellular aggregates remained unaffected.

ALGAL MEGAFOSSILS AND EUKARYOTES

The history of eukaryotes in the pre-Phanerozoic fossil record is controversial. One group of researchers (34) contends that eukaryotes at the microbial level can be recognized by means of one or a combination of several criteria: 1) the presence of preserved internal structures (interpreted as organelles); 2) the juxtaposition of cells (tetrahedral tetrads); 3) cell size (cells >20 μm are candidate eukaryotes, but cells several hundred micrometers in diameter are probably eukaryotic); and 4) morphologic complexity (large diameter, organic walled, branching filaments with cross walls). Other researchers (5,17) contend that it is difficult, if not impossible to establish the eukaryotic nature of any microfossil in the pre-Phanerozoic. Fortunately the geologic record contains less controversial evidence for the existence of eukaryotes in the Proterozoic. Examples are the centimeter-sized, ribbon-like carbonaceous films interpreted to be the remains of algae in 1300 Ma-old rocks (44). The Little Dal biota, ~1000 Ma-old, contains a variety of megascopic, presumably algal, remains (19), and macroscopic spheroids (Chuaria) are thought to be algal as well (19). Probable Chuaria has recently been found in ~1600 Ma-old rocks of China (19). Though rare, these megascopic remains tend to show that eukaryotes were present and already large ~1300 Ma ago.

SUMMARY

The earliest well documented record of life on Earth has been found in the ~3500 Ma-old Warrawoona Group of Western Australia. Here diverse, filamentous microfossils, possible spheroidal microfossils, as well as stromatolites, record the presence of rather advanced microbial organization; these suggest that life is considerably older than the Warrawoona Group. The timing of the appearance of oxygen-releasing photosynthesis, the most significant evolutionary novelty of the biosphere, is uncertain. The similarity of some of the Early Proterozoic stromatolitic microfossils to modern microbes, in particular cyanobacteria as well as the stromatolite constructions themselves, permit one to conclude that cyanobacteria-like microbes were abundant during early Proterozoic time.

The pre-Phanerozoic fossil record is dominated by stromatolites and their preserved microbiotas as well as by abundant acid-resistant microbes that have been isolated from clastic rocks. Prokaryotes dominated most of the pre-Phanerozoic; algae (protists) had appeared ~1400 Ma ago. During the Vendian, columnar stromatolite diversity decreased (presumably in response to metazoan activity), more highly ornamented acritarchs appeared, and robust filaments became more conspicuous in stromatolitic deposits; these changes contributed to making the transition to the Phanerozoic one of the more exciting paleobiological events.

Acknowledgement. I thank B. Mahall, D. Pierce, and D. Schulte for their assistance. Contribution No. 108 of the Preston Cloud Laboratory, UCSB.

REFERENCES

(1) Awramik, S.M. 1971. Precambrian columnar stromatolite diversity: reflection of metazoan appearance. Science 174: 825-827.

(2) Awramik, S.M. 1976. Gunflint stromatolites: microfossil distribution in relation to stromatolite morphology. In Stromatolites, ed. M.R. Walter, pp. 311-320. Amsterdam: Elsevier.

(3) Awramik, S.M., and Barghoorn, E.S. 1977. The Gunflint microbiota. Precambrian Res. 5: 121-142.

(4) Awramik, S.M.; Gebelein, C.D.; and Cloud, P. 1978. Biogeologic relationships of ancient stromatolites and modern analogs. In Environmental Biogeochemistry and Geomicrobiology, ed. W.E. Krumbein, vol. 1, pp. 165-178. Ann Arbor: Ann Arbor Press.

(5) Awramik, S.M.; Golubic, S.; and Barghoorn, E.S. 1972. Blue-green algal cell degradation and its implication for the fossil record. Geol. Soc. Am., Abstr. Progr. 4(7): 438.

(6) Awramik, S.M.; Schopf, J.W.; Walter, M.R.; and Buick, R. 1981. Filamentous fossil bacteria 3.5 x 10^9-years-old from the Archean of Western Australia. Science, in press.

(7) Banerjee, D. 1971. Precambrian stromatolitic phosphorites of Udaipur, Rajasthan, India. Bull. Geol. Soc. Am. 82: 2319-2330.

(8) Barghoorn, E.S., and Tyler, S.A. 1965. Microorganisms from the Gunflint chert. Science 147: 563-577.

(9) Bridgewater, D.; Allaart, J.H.; Schopf, J.W.; Klein, C.; Walter, M.R.; Barghoorn, E.S.; Strother, P.; Knoll, A.H.; and Gorman, B.E. 1980. Microfossil-like objects from the Archaean of Greenland: a cautionary note. Nature 289: 51-53.

(10) Cloud, P. 1976. Beginnings of biospheric evolution and their biogeochemical consequences. Paleobiology 2: 351-387.

(11) Cloud, P. 1976. Major features of crustal evolution. Geol. Soc. S. Afr., Annexure Volume 79: 1-32.

(12) Cloud, P., and Morrison, K. 1979. On microbial contaminants, micropseudofossils, and the oldest records of life. Precambrian Res. 9: 81-91.

(13) Cloud, P., and Semikhatov, M.A. 1969. Proterozoic stromatolite zonation. Am. J. Sci. 267: 1017-1061.

(14) Diver, W.L., and Peat, C.J. 1979. On the interpretation and classification of Precambrian organic-walled microfossils. Geology 7: 401-404.

(15) Downie, C. 1973. Observations on the nature of acritarchs. Palaeontology 16: 239-259.

(16) Dunlop, J.S.R.; Muir, M.D.; Milne, V.A.; and Groves, D.I. 1978. A new microfossil assemblage from the Archaean of Western Australia. Nature 274: 676-678.

(17) Francis, S.; Margulis, L.; and Barghoorn, E.S. 1978. On the experimental silicification of microorganisms II. On the time of appearance of eukaryotic organisms in the fossil record. Precambrian Res. 6: 65-100.

(18) Hofmann, H.J. 1976. Precambrian microflora, Belcher Islands, Canada: significance and systematics. J. Paleontology 50: 1040-1073.

(19) Hofmann, H.J., and Aitken, J.D. 1979. Precambrian biota from the Little Dal Group, Mackenzie Mountains, northwestern Canada. Can. J. Earth Sci. 16: 150-166.

(20) James, H.L. 1978. Subdivision of the Precambrian - A brief review and a report on recent decisions by the Subcommission on Precambrian Stratigraphy. Precambrian Res., 7: 193-204.

(21) Kline, G.L. 1975. Metallogenium-like microorganisms from the Paradise Creek Formation, Australia. Geol. Soc. Am., Abstr. Progr. 7(3): 336.

(22) Knoll, A.H., and Barghoorn, E.S. 1977. Archean microfossils showing cell division from the Swaziland System of South Africa. Science 198: 396-398.

(23) Knoll, A.H.; Barghoorn, E.S.; and Awramik, S.M. 1978. New microorganisms from the Aphebian Gunflint Iron Formation, Canada. J. Paleontology 52: 976-992.

(24) Knoll, A.H.; Blick, N.; and Awramik, S.M. 1981. Stratigraphic and ecologic implications of Late Precambrian microfossils from Utah. Am. J. Sci. 281: 247-263.

(25) Komar, V.A. 1976. Classification of stromatolites according to microstructure. In Paleontology of Precambrian and Early Cambrian. Abstracts of All-Union Symposium, 11-14 May, 1976, Novosibirsk, pp. 41-43. (In Russian).

(26) LaBerge, G.L. 1967. Microfossils and Precambrian iron-formations. Bull. Geol. Soc. Am. 78: 331-342.

(27) Lowe, D.R. 1980. Stromatolites 3,400 M.Y. old from the Archaean of Western Australia. Nature 284: 441-443.

(28) Moorbath, S.; O'Nions, R.K.; and Pankhurst, R.J. 1973. Early Archaean age for the Isua iron-formation, West Greenland. Nature 245: 138-139.

(29) Muir, M.D., and Grant, P.R. 1976. Micropalaeontological evidence from the Onverwacht Group, South Africa. In The Early History of the Earth, ed. B.F. Windley, pp. 595-604. New York: Wiley and Sons.

(30) Oehler, D.Z. 1978. Microflora of the middle Proterozoic Balbirini Dolomite (McArthur Group) of Australia. Alcheringa 2: 269-309.

(31) Schidlowski, J.; Appel, P.W.V.; Eichmann, R.; and Junge, C.E. 1979. Carbon isotope geochemistry of the 3.7 x 10^9 - yr-old Isua sediments, West Greenland: implications for the Archean carbon and oxygen cycles. Geochim. Cosmochim. Acta 43: 189-199.

(32) Schopf, J.W. 1974. Paleobiology of the Precambrian: the age of blue-green algae. In Evolutionary Biology, eds. T. Dobzhansky, M.K. Hecht, and W.C. Steere, vol. 7, pp. 1-43. New York: Plenum Press.

(33) Schopf, J.W. 1977. Biostratigraphic usefulness of stromatolitic Precambrian microbiotas: a preliminary analysis. Precambrian Res. 5: 143-173.

(34) Schopf, J.W., and Oehler, D.Z. 1976. How old are the eukaryotes? Science 193: 47-49.

(35) Schopf, J.W., and Sovietov, Y.K. 1976. Microfossils in Conophyton from the Soviet Union and their bearing on Precambrian biostratigraphy. Science 193: 143-146.

(36) Schopf, J.W., and Walter, M.R. 1980. Archaean microfossils and "microfossil-like" objects - a critical appraisal. In Extended Abstracts, Second International Archaean Symposium, eds. J.E. Glover and D.I. Groves, p. 23. Perth: Geological Society of Australia.

(37) Semikhatov, M.A. 1976. Experience in stromatolite studies in the U.S.S.R. In Stromatolites, ed. M.R. Walter, pp. 337-357. Amsterdam: Elsevier.

(38) Semikhatov, M.A.; Gebelein, C.D.; Cloud, P.; Awramik, S.M.; and Benmore, W.C. 1979. Stromatolite morphogenesis - progress and problems. Can. J. Earth Sci. 16: 992-1015.

(39) Serebryakov, S.N. 1975. Peculiarities of formation and location of Riphean Siberian stromatolites. Trans. U.S.S.R. Acad. Sci. Moscow 200: 175. (In Russian).

(40) Timofeev, B.V. 1973. Plant microfossils from the Proterozoic and lower Paleozoic. In Microfossils of the Oldest Deposits: Proceedings of the 3rd Intern. Palynol. Conf., ed. T.F. Vozzhennikova and B.V. Timofeev, pp. 7-12. Moscow: Nauka. (In Russian).

(41) Vidal, G. 1976. Late Precambrian microfossils from the Visingsö Beds in southern Sweden. Fossils and Strata 9: 1-57.

(42) Walter, M.R. 1980. Archean stromatolites: the history of the Earth's earliest benthos. In A Symposium: Interdisciplinary Study of the Origin and Evolution of Earth's Earliest Biosphere, U.C.L.A., Los Angeles, CA, Abstracts, pp. 10-11.

(43) Walter, M.R.; Buick, R.; and Dunlop, J.S.R. 1980. Stromatolites 3.4-3.5 billion years old from the North Pole area, Pilbara Block, Western Australia. Nature 441: 443-445.

(44) Walter, M.R.; Oehler, J.H.; and Oehler, D.Z. 1976. Megascopic algae 1300 million years old from the Belt Supergroup, Montana: a reinterpretation of Walcott's Helminthoidichnites. J. Paleontol. 50: 872-881.

(45) Zhou, H., and Han, Y. 1973. On the textural and structural characteristics of the Yingmeng iron-ore deposit near Beijing and a preliminary discussion of the genesis of the deposit. Peking Geol. 1: 26-40. (In Chinese).

Mineral Deposits and the Evolution of the Biosphere, eds. H.D. Holland and M. Schidlowski, pp. 83-102. Dahlem Konferenzen, 1982.
Berlin, Heidelberg, New York: Springer-Verlag.

Precambrian Evolutionary Genetics

J. Langridge
Commonwealth Scientific and Industrial Research Organization
Canberra City, A. C. T. 2601, Australia

Abstract. The broad features of the evolution of Precambrian organisms and of the genetic systems responsible for this evolution are outlined. The hypothetical and actual life-forms considered are pre-organisms (nucleic acid molecules multiplying free in solution), proto-organisms (membrane-bound self-replicating entities), prokaryotes (bacteria and blue-green algae), and protists (early eukaryotes with little differentiation). In each group, the nature of the hereditary material, its acquisition of function, its increase in size and variety, and its means of change and recombination are discussed.

INTRODUCTION

Many biologists believe that the three thousand million or so years of life in the Precambrian are of little interest. The remains of such organisms as may have existed are confusing; nothing can be known of the manner in which they functioned, and their relevance to familiar forms of life is considered dubious. However, simple reflection on the nature of evolutionary origins leads one to the conclusion that life originated in the Precambrian, and that most of the inventions that condition life were evolved by Precambrian organisms. Some organisms differ little from their Precambrian ancestors; these are the prokaryotes (bacteria and blue-green algae) as well as the unicellular and the simply differentiated multicellular eukaryotes. One can arrive at an approximation of the early condition of these organisms by recognizing primitive features, by simplifying their present structure and functions, and by extrapolating known evolutionary trends. The same procedure, combined with a knowledge of ancient environments and the application of chemical principles, can be used to infer the nature of the extinct entities which must have preceded the prokaryote-like forms. This approach has been reasonably successful in defining the probable metabolic sequences that maintained the early forms of life, but there has been little discussion of their genetic aspects. Such discussions are undoubtedly more speculative than those dealing with metabolism but they are equally important. A treatment of genetics is inadequate but does seem worth putting foward now, because the field of molecular biology is providing information relevant to the subject.

In a short communication it is not possible to consider all of the evolutionary genetic features of Precambrian life; particular topics have therefore been chosen for discussion. A rather general view, in the evolutionary sense, of the transition from the first forms of life to the beginnings of higher organisms, is followed by a treatment of early genetic systems, of the manner in which genetic information has accumulated, and of the role of evolution in expanding and refining certain inherent properties of matter.

THE COURSE OF PRECAMBRIAN EVOLUTION

It is convenient to divide the hypothetical forerunners of the prokaryotes into pre-organisms and proto-organisms. Pre-organisms were probably small nucleic acid polymers; whether these consisted of RNA or DNA is uncertain. RNA molecules can be multipled in vitro; the simplest known natural self-replicating nucleic acids are the small, single-stranded RNA molecules of the viroids. However, the linkages between subunits in DNA are about 10 times more stable than those in RNA, and DNA is less sensitive to radiation damage. For these reasons, it is generally thought that "life" began as randomly polymerized circular RNA molecules; these are thought to have been replaced at some unspecified later stage by more stable DNA polymers.

The proto-organisms are more advanced; their nucleic acid was enclosed in a semi-permeable boundary membrane composed of a layer of lipid, protein, or lipoprotein. The concentration of high molecular weight compounds within the membrane led to the evolution of a template influence or specification on amino-acid polymerization, which in turn provided catalysts for replication, energy transformations, and protection against radiation damage.

These, or similar, very primitive organisms were perhaps superseded more than $3x10^9$ years ago by the prokaryotes with their highly evolved systems for the control of genetic, catalytic, and synthetic functions. Such systems are probably present in all prokaryotes, and it is generally not possible to trace their evolution from extant simple to advanced organisms. The blue-green algae may be an exception, because certain of their usual biochemical syntheses seem to be poorly regulated if at all (2,12). It is logical, therefore, to suppose that the cell itself and its basic biochemistry were evolved in the extinct precursors of the prokaryotes.

Furthermore, it seems reasonable that proto-organisms were replaced by prokaryotes similar to the existing genera by virtue of the evolution in the latter of means of fine control of the living processes built up by the proto-organisms. The adaptive advantages conferred by precise and environmentally sensitive control systems include the conservation of energy and carbon compounds, the prevention of unwanted side-reactions, and the attainment of rapid and complete reproduction.

The evolution of high efficiency in biological functioning may have been forced on the predecessors of the present prokaryotes by the origin of the eukaryotes in the middle or late Precambrian (10). If this is correct, the evolution of the typical contemporary prokaryotic cell and the extinction of its proto-organism precursor coincided with the origin of the eukaryotic cell.

The control mechanisms of prokaryotes are variants of a few simple strategies based on interactions between different proteins, between proteins and nucleic acids, and between proteins and small organic compounds. Nevertheless these controlling activites may require half the genes of even the simplest prokaryote.

The size and general morphology of some contemporary prokaryotes are very like those from the Fig Tree cherts with an age of about $3x10^9$ years; this has led to the belief that prokaryotic evolution has long been arrested. There are, however, quite marked differences in DNA sequences even between closely related genera of bacteria. In fact, there is more diversity in the genetic nucleic acids of certain species of bacteria than in certain mammalian genera (5,9). It is not, however, certain how much of this divergence in bacterial nucleic acids has adaptive significance. Some nucleotide substitutions were probably concerned with fine aspects of differential adaptation, but the majority are probably selectively neutral differences which have accumulated since early evolutionary divergence.

There appear to be two parts to the genetic constitution of modern prokaryotes. The first comprises the single, large DNA molecule that is the equivalent of a eukaryotic chromosome; it contains all of the genes necessary for growth, energetic and synthetic metabolism, and reproduction. These genes have been evolved for a long time; they do not seem to vary much from one bacterium to another and may not be very different from those of their Precambrian ancestors. It is likely that at present, and for some considerable time, prokaryotic evolution has been based largely on plasmid formation, change, and transfer. The plasmids are small, independently replicating, double-stranded DNA circles which are separate from the main genetic DNA. They carry genes for adaptation to specific aspects of the environment such as for resistance to antibiotics and for the use of unusual carbon sources. However, these plasmid genes may not have been present in the Precambrian organisms because the nature of their adaptive advantage is rather special. It seems unlikely that they contributed to later eukaryotic evolution.

One of the largest gaps in the evolutionary succession of organisms exists between the prokaryotes and the eukaryotes. With the exception of viruses, all organisms possess one of only two general patterns of subcellular organization, the simpler pattern characterizing the prokaryotes and the more complex, the eukaryotes.

Among living organisms, those closest to the form expected for the earliest eukaryotes are found in the large, heterogeneous group sometimes called the Protista. Its members (algae, protozoa, slime molds, and fungi) are lower eukaryotic organisms, which have a wide range in level of organization above those of the prokaryotes, but below those of vascular plants and animals. Typically they lack tissues, and differentiation is mainly confined to their reproductive structures. Many of them are haploid (only a single gene of each sort) and have generation times of only a few hours; others are dipolid (two sets of genes), sexual organisms with long generation times. Consequently, the contribution to the evolution of mutation, recombination, and segregation varies widely within different groups of Protista. The outstanding contribution of the Protista to the course of evolution is their development of two novel features. The first is the acquisition of the intracellular organelles, which are responsible for aerobic respiration and for photosynthesis; the second is their system of mitosis for transmitting genetic information to succeeding cell generations. Certain innovations in body organization which have been especially important in setting directions in plant evolution are more sporadic in the Protista.

The weight of present evidence is in favor of the derivation of chloroplasts, and perhaps of mitochondria, as well as from small prokaryotic organisms that were permanently included in a larger host cell (8). The original organisms forming these combinations cannot reliably be identified with any existing prokaryotes which, on the reasoning presented earlier, may be because they were the now extinct predecessors of the prokaryotes. On the other hand, a free-living mycoplasma, Thermoplasma acidophilum, has been described which has certain of the characteristics expected for the original host (8). It is non-photosynthetic and has a relatively inefficient energy metabolism because it apparently lacks both cytochrome-type respiration and oxidative phosphorylation (11). If this organism is representative of pre-eukaryotic life, the large increase in energy efficiency brought about by the symbioses would cause intense competition with the existing organisms and a rapid evolution of the proto-organisms or early prokaryotes to the status of typical prokaryotes. The other characteristic feature of the eukaryotic cell, mitosis, presumably evolved following the endosymbiotic events, because different intermediate stages in its structure and function are seen in some simple organisms with chloroplasts and mitochondria.

The prokaryotes never achieved any but the simplest forms of differentiation despite occasional multicellularity, and their cells remained small and practically undifferentiated. Their simplicity has variously been attributed to a primitive gene structure, to inadequacies in regulation or, to relatively poor energy transformation systems in the prokaryotes. Whatever the reason, differentiation remains largely a property of the later-evolved eukaryotes. There it is distinguishable at two levels. The first is in single-celled organisms where some of the unicellular Protista, the predatory ciliates, have achieved a cell complexity that is unequalled in any other organism. All ciliates have special organs for movement, mouths, and anuses; some have an esophagus-like tube, a rectum-like tube, and something resembling an alimentary canal, all in a single cell. This is probably due to a special genetic device in the Ciliata (4) and in a few Foraminifera and is not an intermediate state of progressive evolution. The second type of differentiation is based on separate genetic programs to form different types of cells and requires multicellularity. Multicellularity can readily evolve and has done so repeatedly, but by and of itself it is probably not sufficient to set the stage for cellular differentiation. There also needs to be a certain level of integration or cooperativity of an organism's component cells. This is suggested by a gradation towards integration and an increase in differentiation in some of the multicellular green algae (7).

THE ACQUISITION OF INHERITED INFORMATION

The substance of the pre-organisms probably consisted of randomly polymerized nucleic acid chains; these contained little information except perhaps the ability to bind accessory inorganic or organic compounds which aided their replication or enhanced their stability. Amino acid chains were probably among those catalytic compounds; they were presumably short in sequence and variable and unstable in structure. Their polymerization must have been influenced by the nature of the nucleic acid. Such an association could have provided the first step towards a nucleic acid-protein coding system and thus could have added an informational role to the replicative role of the nucleic acids. This type of direct interaction between nucleic acids and proteins is very different from that operating today, and its real nature is unknown. However, its further evolution probably required the presence of a boundary membrane separating the replicating entities from the surrounding dilute aqueous medium. Since some spontaneously polymerized polypeptides have catalytic activity, even a weak ordering influence on amino acid polymerization would give certain nucleic acid chains a significant advantage. The order of acquisition of genetic function, i.e., nucleic acid sequences specifying catalytic polypeptides is expected to be, first, a transferase catalyzing group exchange in the polymerization of nucleotides and amino-acids; second, an oxidase capable of removing hydrogen from organic compounds, of transferring it to a hydrogen acceptor, and of using the energy so derived to couple phosphate ions; and third, an enzyme-catalyzed mechanism to repair ultraviolet-induced damage to the nucleic acids. The pre-enzymes would have had rather little specificity; even today enzyme specificity is frequently alterable by mutation. Basic catalytic activities are more referable to the type of bond that is rearranged; enzymes may therefore be classified as transferases, hydrolases, oxidoreductases, lyases, ligases, and isomerases. These six activities represent the maximum number of unique gene sequences that would need to be evolved in proto-organisms. Actually, the number was probably fewer, because lyases, ligases, and isomerases perform rather specialized catalytic activities which could have evolved later by very rare mutational changes to new forms of bond rearrangement. Genes must have increased greatly in size during the early Precambrian. The average size of a present-day gene is 1,000 nucleotides. The probability of its formation by random polymerization is about 10^{600}. Clearly, any particular current enzyme coded for by a gene must have been preceded by a very much smaller, less efficient molecule. For example, a mammalian esterase capable of hydrolyzing p-nitrophenyl-caproate contains over 400 amino acids, yet we have been able to make a polypeptide that catalyzes this hydrolysis in an enzymatic fashion and contains only five amino acids (J. Langridge and K. Bentley, unpublished). The genes of the higher eukaryotes are about 25 per cent larger than those of bacteria (6), but the mechanism by which their size increased is unknown.

An examination of even very primitive extant organisms shows that in the course of gene evolution, one or a few genes specifying enzymes of little specificity and low catalytic activity have been replaced by a series of descendant genes specifying enzymes of progressively increasing specificity and catalytic efficiency. While nucleic acid was merely a self-replicating form of matter with little or no information content, it could increase in size by haphazard means such as "slippage," where mistakes in replication repeated part of the nucleotide sequence. With the advent of transmissible information in terms of an ordering influence on amino acid polymerization, a meaningful increase in molecular size could only occur by partial or complete duplication of existing molecules. It is well-known that both processes have occurred in the DNA of numerous organisms, and that they are possible because the phospho-diester bonds are relatively labile.

It can be shown that the number of reactions involved in the growth and metabolism of an organism requires about 500 genes for life on a simple preformed carbon source. But the bacterium with the smallest known amount of DNA, Mycoplasma meleagridis, has about 1,000 genes. It seems, therefore, that a simple unicellular organism needs roughly as much genetic information for control as it does for synthetic, degradative and energy-yielding mechanisms. This great increase in gene number over the rather few postulated to have existed in early pre-organisms may not have occurred gene by gene but by successive doublings of the total nucleic acid; this process has been demonstrated in some lines of bacterial evolution. Wallace and Morowitz (13) have proposed that organisms like the contemporary Mycoplasma were the ancestors of the prokaryotes. They base this proposal on small genomes (DNA contents), small cells, and other primitive features. They consider that a doubling of the size of the genome would give rise to forms like the Acholeplasma. These are relatives of the Mycoplasma, which also lack cell walls but appear to have a greater ability to synthesize. A further increase in genome size and the acquisition of cell walls would give the range typical of modern bacteria. This process of deriving new genes has occurred repeatedly. The bacterial-type genome has been multiplied more than 700 times to give about 4,000,000 genes or gene equivalents in the nuclear DNA of man.

EVOLUTION OF THE GENETIC APPARATUS

The genetic system of an organism comprises the means of transmitting information coded in nucleic acids to progeny molecules or to cells and the means of altering or adding to this information during the course of evolution. The processes involve nucleic acid duplication, deletion, mutation, combination with the nucleic acid of another organism, and orderly segregation of the nucleic acid to the progeny.

The pre-organisms merely contained certain large polymers which could be multiplied. The transmission of this "animate"; type of matter to succeeding entities is the genetic aspect, and its inherited change with time is the evolutionary aspect.

Organic compounds that have been experimentally synthesized under reducing conditions commonly include only three types of molecules capable of polymerization: nucleic acid bases, amino acids, and sugars. With the types of amino acids then available, it seems unlikely that a protein capable of replication via pairing of opposite charges could have formed. The known polysaccharide chains do not have the structure necessary for linear copying on a template basis. On the other hand, nucleic acid bases, which are planar molecules because of their resonance characteristics, have amino- and keto-groups appropriately situated for complementary hydrogen bonding. A further reason for regarding the nucleic acids rather than the proteins or the polysaccharides as the initial form of replicating matter is that the former have the capacity for spontaneous change, and thus evolution built into them. The purines and pyrimidines both have a very stable resonating structure to which are attached potentially reactive groups that can exist in relatively long-lived excited states. The existence of these rare tautomeric forms of the bases is probably the main cause of spontaneous mutation.

During self-replication one molecule acts in an autocatalytic fashion to produce another molecule nearly identical with itself. In such a system, any spontaneous changes in the linear sequence of bases present at the time of replication will be passed on to progeny molecules. These changes, which nucleic acid bases are prone to undergo, may increase or decrease the rate of further molecular reproduction. Consequently, the molecular population will be diverse, and in time those that can reproduce most rapidly and accurately will dominate the population. This is comparable with known processes of evolution by mutation and selection where, at an organismal level, those that multiply most readily and most precisely eventually predominate. Whereas the mass ratio of animate to inanimate matter initially increased merely because the former was able to replicate, a pool of replicating molecules of different structures existed later, and competition developed between them.

The development of proto-organisms with a membrane surrounding the nucleic acid presented a special problem, because in these organisms the distribution of the nucleic acid must have been geared to the division of the cell. The spherical membranes of the first proto-organisms probably enclosed a number of nucleic acid molecules and, as the sphere split into two or more smaller spheres, these molecules were distributed passively and at random. This procedure is very wasteful of nucleic acid and must soon have been replaced by the special segregation mechanism that is still in use. The method used by all known bacteria is to attach the cell's DNA to a particular place on the inner surface of the membrane. Once the DNA has been duplicated, the region of the membrane between the duplicates expands, and a new membrane is formed in the expanded region to give two cells, each with an identical molecule of DNA.

The proto-organisms probably acquired additional nucleic acid polymers either by the accidental fusion of spheres or by simple absorption. The latter process is comparable to the process of transformation in bacteria; exogenous DNA is taken up and is incorporated into the DNA of the host by enzymatic steps of breakage and joining (recombination).

A mechanism for recombining DNA probably evolved either in the proto-organisms or in the early prokaryotes. The incidence of ultraviolet light at that time was probably much higher than it is now; even today it causes a linking between adjacent pyrimidines at a rate of about ten per minute in a bacterial chromosome. Since none of the known DNA replicating enzymes can synthesize DNA past such a "dimer," these events are potentially lethal. There was therefore, an early requirement for a process to repair DNA. A potentially ancient repair system of this sort is the splitting of pyrimidine dimers by an enzyme using absorbed light energy. More recent repair mechanisms excise the damaged DNA bases and fill the gaps with newly synthesized DNA. This repair system, first evolved as protection against ultraviolet light damage, may have been developed further for recombination or gene exchange. The interrelation of ultraviolet light effects and recombination is still evident in contemporary bacteria, in which ultraviolet light stimulates recombination and recombination restores DNA that has been damaged by ultraviolet light. The advantage of recombination between individuals possessing a different favorable mutation is that the chance of incorporating both mutations in one individual is at least twice that of obtaining them by successive mutations. Therefore, rates of production of new gene combinations in the whole genome, and thus the potential rates of evolution, are hundreds of times greater with recombination than with sequential mutation.

Present-day prokaryotes may have either or both of two further means of gene exchange in addition to transformation. One (transduction) relied on the carriage of prokaryotic genes by bacterial viruses from one individual to another; however, these specialized viruses probably evolved late. The other prokaryotic mechanism of gene exchange (sexual reproduction) is also thought to be relatively recent, since it may be a secondary consequence of the system used by plasmids for their transfer between cells.

After the evolution of the eukaryote cell type, the methods of segregating duplicated DNA in the dividing cell changed notably. All but a few of the simplest eukaryotes have a special structure, the spindle, and a special mechanism, mitosis, to separate the DNA during cell division. However, the presence of segregation mechanisms reminiscent of the prokaryotic type in certain algae suggests that the mitotic system for DNA separation may not have been present until late in the Precambrian.

Above a certain size, a single genetic molecule of the bacterial type is likely to be difficult to duplicate and segregate entirely into different cells. Thus it became necessary to distribute genes into separate chromosomes; this process occurred uniquely in all known eukaryotes. The appearance of repeated nucleotide sequences, various patterns of sequence organization, and protein-DNA associations that were not present in the prokaryotes are correlated with the presence of chromosomes. These novel features of the genetic material are frequently supposed to be concerned in the origin or control of development, but evidence for their role is lacking.

Sexuality is a widespread property of the eukaryotes: it involves the ability of one or more organisms to combine their genetic information and usually to redistribute it subsequently to their progeny. Eukaryotic sexuality bears little resemblance to the rare, functionally equivalent processes of prokaryotes. Unlike the prokaryotic mechanism, each sexual event in eukaryotes results in a doubling of the genome, a matter which has required the evolution of a corrective process. The system for the reduction of the genetic material, meiosis, appears to have evolved from mitosis and is another novel feature of most eukaryotes.

THE DOCTRINE OF EMERGENCE

We continually re-encounter the question of the boundary between life and matter. We ask whether life is a manifestation of matter at a certain level of organization or whether it is the result of an interaction between matter and an unspecified non-material principle. The philosophical view is that the latter is the more probable, because there is more in the result (of evolution) than there is in the cause. The doctrine of emergence requires a life principle to be involved, because an explanation is thought to be impossible in physicochemical terms alone. The philosophic argument can only be countered by showing that the characteristics of life are merely realizations of properties which were latent in the purely material beginnings, and that evolution is the agent of their realization.

Except for the properties peculiar to the carbon atom, which make it essential to life, the chemical features amplified and refined by evolution belong to the high polymers, nucleic acids, and proteins. As we have seen, it is only the "fortuitous" structure of the purines and pyrimidines which are capable of linear joining so as to leave exposed amino- and keto-groups that allows a form of multiplication of matter. Yet these subunits can be made by non-biological means, can be polymerized, and caused to multiply in vitro (1). Although the polymerizations have so far only been done with a biological protein catalyst, it seems likely that completely synthetic catalysts could be used.

The hydrophilic nature of the nucleic acids is necessary for solubility in water; this gives them the capacity to combine with certain other molecules. Since the bases themselves are uncharged under ordinary conditions, it is the deprotonated hydroxyl groups of the phosphodiester linkage that provide sites for association with basic substances. Such associations might be expected to evolve if they could advantageously alter the structure and function of the nucleic acid. In fact, the genetic nucleic acids of all organisms except the viroids are complexed with proteins. As far as is known, the main use of the nucleic acid-protein interaction is to control the expression of genes. It is not only desirable to be able to direct available energy, precursors, space, and solvent capacity to meet the immediate needs of the organism; a time-dependent release of genetic information is also required for much of differentiation. The former is provided by protein repressors and their variants, the latter probably by changes in basic and acidic proteins associated with the nucleic acid. Of course, these are late-evolved interactions, for the protein must bind specifically and carry within its structure the basis of a mechanism operating its release from its nucleic acid site.

Apart from their combination with nucleic acids to provide controls in gene expression and their role in stabilizing nucleic acid polymers, proteins are most important in providing catalysts for organic reactions. Experiments suggest that sequences of artificially synthesized polypeptides are not random, but that they have a pattern of adjacent amino acids rather like that in natural proteins. Some of these simple polymers are able to catalyze a number of bond rearrangements. They exercise the greatest catalytic effect on hydrolytic reactions, transferring the hydroxyl group and hydrogen atom of water to the two products of the reaction, but decarboxylation, amination, and deamination have also been demonstrated. These catalytic activities conform to the kinetic rules of enzyme action and in two cases, the hydrolysis of p-nitrophenyl acetate and the decarboxylation of oxaloacetic acid, the catalyst appears to be regenerated after reaction. Thus, even the most important feature of catalysis may have been latent in the pre-biological starting matter.

Certain of the spontaneously-synthesized polypeptides appear to contain varying amounts of α-helix. This is the most common of three types of secondary structures found in biological proteins. In the α-helix hydrogen bonding between carbonyl oxygens and amide hydrogens three amino acids back in the polypeptide chain gives a coil structure and shortens the polypeptide. Its significance in evolution is that there is evidence that muscular contraction rests on a transition between the helical and random coil configurations (3). By shortening or lengthening the protein molecules of myosin in the fibers of muscle cells, the muscular tissue can be caused to contract or lengthen. Consequently, a wide variety of movements of parts of the organism and of the organism itself becomes possible. This is a further example of the manner in which very complex functions, and indeed a whole avenue of evolution, is derived from a fundamental property of the abiological matter used in forming living systems. Amino acid polymers, whether artificially or biologically synthesized, have a further inherent property which has been especially important in many lines of evolution. The presence in the protein of even relatively short sequences of hydrophobic amino acids implies that in an aqueous solvent the proteins will tend to aggregate via these sequences. This property was probably present in pre-biological polypeptides where the relatively high concentration of hydrophobic aliphatic amino acids would have favored solvent repulsion and thus aggregation because of their enhanced stability. This tendency for aggregation has been modified and amplified during evolution to provide enzymes with associated control proteins, microtubules and microfilaments, axial fibers, flagella, cell associations, etc. It has become a basic determinant in morphogenesis at the subcellular, cellular, and tissue levels.

As mentioned earlier, when an unspecific protein is mixed with lipid, there is a lateral association into membrane-like sheets. Artificially-formed membranes and those whose formation is under biological control are semi-permeable. Consequently, some monovalent ions pass readily into the sphere or cell while others tend to remain outside, generating an electric potential across the membrane. A disturbance of this "resting" potential as, for example, by contact with a foreign body, may cause a change in ionic permeability and thus a flow of ions. The later evolution of cells specifically designed to detect these changes in current laid the basis for a nervous system. This has evolved weakly if at all in the immobile plants; because of its importance in controlling movement, it has become very highly evolved in most of the multicellular animals, at least from the Coelenterata onwards, and has culminated in the brain and consciousness of man.

As more biological phenomena become interpretable at the molecular level, they are likely to be understood in terms of reactions and interactions that are intrinsic to the organic starting matter. Even now it can be argued that such fundamental properties of the living state as self-multiplication, catalysis of reactions, shape and form, movement and environmental response are simply realizations of the potential organization of randomly moving atoms of the atomic cloud that formed the earth. The evidence presented here does not support the doctrine of emergence.

CONCLUSION

In spite of the evidence advanced above against the doctrine of emergence, evolution - even in the Precambrian - gives the appearance of being oriented. In higher organisms, it is evident that some evolution has taken place in particular directions and not in others that seem to be equally possible. The sustained trends that may occur in the development of a particular organ or feature, the progressive reduction with time in certain other characters and the occurrence of parallel and convergent evolution are cases in point. These are relatively small-scale trends that are difficult to define in Precambrian organisms. The large-scale evolutionary direction over all geological periods is the tendency toward ever-increasing complexity. This implies that the more recent organisms usually have a greater number of facets to their organization, and that these facets are more diverse. This directionality is shown by a progression with time in the amount of genetic information, in the amplification of weak initial properties, in the development of systems of gene change and exchange, and in size, structure and variety. The drive towards more advanced organization must rest, since it is a manifestation of the action of the genetic material, on increases in the complexity of the hereditary nucleic acids themselves. These are known to be unstable molecules which undergo spontaneous additions, deletions, and changes in sequence. Therefore it might be said that evolutionary progress is a property inherent in the nucleic acid chain. Especially in the Precambrian, the basic component of evolutionary advance must have been the partial or total duplication of existing nucleic acid. But such duplications merely give an increased synthetic and energetic load without an increase in information or complexity, until they are later altered by mutation. Perhaps, in the absence of a variety of life forms, selective forces were much weaker then or perhaps, at that time, some duplications by themselves were temporarily advantageous. At this level there seems to be some component of the Precambrian evolutionary process which eludes our understanding. Evidently, the mechanism by which evolution takes the path to higher complexity is rather more unclear than the reason for this path, which is that an increase in the complexity of structure or function implies accentuation or addition and tends to endow organisms with competitive advantages.

REFERENCES

(1) Biebricher, C.K., and Orgel, L.E. 1973. An RNA that multiplies indefinitely with DNA-dependent RNA polymerase: selection from a random copolymer. Proc. Nat. Acad. Sci. USA 70: 934-938.

(2) Carr, N.G. 1973. The Biology of the Blue-Green Algae, eds. N.G. Carr and B.A. Whitton, pp. 39-65. Blackwell: Oxford.

(3) Harrington, W.F. 1979. On the origin of the contractile force in skeletal muscle. Proc. Nat. Acad. Sci. USA 76: 5066-5070.

(4) Iwamura, Y.; Sakai, M.; Mita, T.; and Matsumura, M. 1979. Unequal gene amplification and transcription in the macronucleus of Tetrahymena pyriformis. Biochemistry 18: 5289-5294.

(5) Jones, D., and Sneath, P.H.A. 1970. Genetic tranfer and bacterial taxonomy. Bacteriol. Rev. 34: 40-81.

(6) Kiehn, E.D., and Holland, J.J. 1970. Size distribution of polypeptide chains in cells. Nature 226: 544-545.

(7) Kühn, A. 1971. Lectures in Developmental Physiology, pp. 112-118. New York: Springer-Verlag.

(8) Margulis, L. 1970. Origin of Eukaryotic Cells. New Haven: Yale University Press.

(9) Moore, R.L., and Hirsch, P. 1972. Deoxyribonucleic acid base sequence homologies of some budding and prosthecate bacteria. J. Bacteriol. 110: 256-261.

(10) Schopf, J.W.; Haugh, B.N.; Molnar, R.E.; and Satterthwaite, D.F. 1973. On the development of metaphytes and metazoans. J. Paleontol. 47: 1-9.

(11) Searcy, D.G.; Stein, D.B.; and Green, G.R. 1978. Phylogenetic affinities between eukaryotic cells and a thermophilic mycoplasma. BioSystems 10: 19-28.

(12) Singer, R.A., and Doolittle, W.F. 1975. Control of gene expression in blue-green algae. Nature 253: 650-651.

(13) Wallace, D.C., and Morowitz, H.J. 1973. Genome size and evolution. Chromosoma 40: 121-126.

Mineral Deposits and the Evolution of the Biosphere, eds. H.D. Holland and M. Schidlowski, pp. 103-122. Dahlem Konferenzen, 1982.
Berlin, Heidelberg, New York: Springer-Verlag.

Content and Isotopic Composition of Reduced Carbon in Sediments

M. Schidlowski
Max-Planck-Institut für Chemie, 6500 Mainz, F. R. Germany

Abstract. Reduced carbon, commonly a residuum of biological activity, has been a conspicuous constituent of sedimentary rocks since the start of the rock record 3.8 x 10^9 yr ago. The 36,000 C_{org} assays presently available for Phanerozoic sediments indicate that the average organic carbon content of sedimentary rocks has oscillated around a mean of 0.5 - 0.6% during the last 600 million years. Lack of detailed correlation between observed C_{org} variations and the isotope age curve of Phanerozoic carbonates raises doubt as to whether all of the variations reported are real. However, the higher organic carbon content of Carboniferous and younger rocks seems to be reflected by more positive levels of the $\delta^{13}C_{carb}$ which is to be expected from mass balance considerations. C_{org} assays for Precambrian rocks fall within the scatter of the Phanerozoic data, suggesting that the organic carbon content of Precambrian sediments does not differ significantly from that of geologically younger formations. The constancy of the isotopic fractionation observed between reduced and oxidized carbon throughout the record is best interpreted as the signature of biological activity during the past 3.5 x 10^9 yr (or possibly 3.8 x 10^9 yr).

REDUCED CARBON IN SEDIMENTS: GENERAL BACKGROUND

Carbon is stored in the Earth's sedimentary shell as either reduced (C_{org}) or oxidized carbon (C_{carb}). It is commonly accepted that almost all of the reduced carbon is ultimately derived from living organisms and the products of their metabolism. The primary source of this reduced carbon is, therefore, photosynthetic activity by bacteria and plants since conversion of inorganic carbon into organic carbon is, for the most part, the result of photosynthetic carbon fixation. According to recent estimates, production rates in the biological carbon cycle are on the order of some 10^{16}g C_{org}/yr; net primary productivity on the continents is believed to exceed the primary productivity of the seas (4,17,44,47). These rates are sustained by a standing biomass approaching 10^{18}g carbon. Almost all organic matter produced in this biological cycle is ultimately either recycled by other organisms which use it as a source of both carbon and energy, or destroyed by physicochemical processes. However, between 10^{-2} and 10^{-3} of the total annual turnover rate (17) is trapped in newly-formed sediments. This "leak" has, over the ages, produced a sedimentary reservoir of some 10^{22}g of organic carbon, surpassing the amount of carbon fixed in the stationary living biomass by at least 4 orders of magnitude.

The organic fraction of sediments undergoes a series of transformations during and shortly after burial. Decomposition by microbial degraders, notably sulfate reducers which browse on the energy-rich substrates, releases CO_2 and substantially reduces the quantity of organic carbon ending up in sedimentary rocks. The main components of living matter (carbohydrates, proteins, lipids, lignin) are broken down to their monomers and other simple molecules which, in turn, interreact to give rise to reconstituted compounds of a more complex type (10). These processes, often summarized as "humification," result in the formation of a new generation of polycondensed organic compounds ("geopolymers") which, strictly speaking, do not belong to the living realm. Only a small portion of the original breakdown products is able to survive these processes. Organic molecules that are particularly resistant to degradation and hence most likely to survive diagenesis and reconstitution are aptly referred to as chemofossils or "biological markers" (11).

The organic constituents of sediments are, therefore, parts of a complex, dynamic chemical system. The end product of the evolution of the organic carbon compounds in this system is kerogen (10), the most abundant form of organic matter in the sedimentary shell (and on Earth). Kerogen is defined as the fraction of diagenetically stabilized sedimentary organic matter which is insoluble in the usual organic solvents (as opposed to soluble "bitumens"). Its chemical inertness reflects its highly polymerized or polycondensed state. Kerogen displays a considerable degree of stability under a variety of geological conditions, thus constituting the geopolymer par excellence. The maturation of kerogenous material is characterized by increasing aromaticity and a progressive removal of hydrogen and oxygen; dehydrogenation occurs, in part, by the loss of H-rich organic molecules and may lead to the formation of economic accumulations of hydrocarbons in the sedimentary shell (10,38). The maturation pathways of various protokerogens converge in the formation of a highly polycondensed variety of kerogen that is very poor in hydrogen and oxygen. Graphitization of this end member of the maturation series may start in the upper greenschist facies and is pronounced in amphibolite-grade rocks.

THE QUANTITY OF SEDIMENTARY ORGANIC MATTER AND ITS VARIATION WITH TIME

The quantity of kerogen in a sediment is usually assessed on the basis of its content of reduced or organic carbon (C_{org}). The C_{org} content always gives a lower limit for sedimentary organics; it must be multiplied by a factor of 1.5 - 1.6 for immature kerogens and by 1.2 - 1.3 for the more mature species to obtain the concentration of total organics.

Reliable information on the C_{org} burden of common sediments has accrued during the last few decades, principally as a result of measurements by Trask and Patnode (39) and by Ronov (29). The data base of the classical survey by Ronov on the sedimentary cover of the Russian Platform comprised altogether 25,742 samples grouped into 1,105 composites, 418 of which represented clays, 412 sandstones, and 285 carbonates. The work by Trask and Patnode (39) on sediments from North America was based on 1,052 samples, divided less elaborately into "clastic rocks" and "limestones."

A synopsis of these data is presented in Fig. 1 where the C_{org} contents of shales and average sediments (shale + carbonate + sandstone) are plotted as a function of time. While Phanerozoic shales average 0.67% C_{org}, the means for all sediments of the North American and Russian Platforms are 1.08 and 0.40% respectively. According to Ronov (29), the higher mean yielded by the North American rocks is likely to reflect an overrepresentation in the sampled suite of sediments from oil-bearing areas; the Russian mean was claimed to be more representative of average sediments. In both surveys, argillites contain the bulk of the sedimentary organics (usually between 0.2 and 1.6% C_{org}, with a mean close to 0.7), while carbonates and arenites average about 0.2% C_{org}. In a critical appraisal of a follow-up of the previous work of the Russian school (31), Hunt (18) has concluded that the C_{org} values reported for continental shales by the Russian authors were somewhat low. Using Hunt's revised figures for the C_{org} content of the principal sediment types grouped by provenance (continental plus shelf vs. oceanic), we arrive at calculated C_{org} means for total shales, carbonates, and sandstones of 0.77, 0.30, and 0.27% respectively. Weighted by the relative abundance of the host rocks (Table 3 in (18)), the value for the C_{org} content of the average sediment is 0.54%. In a latest synopsis of the chemical composition of the sedimentary shell, Ronov (30) has proposed values of 0.92, 0.33, and 0.37% for shales, carbonates, and sandstones of continental platforms as well as a mean of 0.62% for the average sediment. If we take the mass of all sedimentary rocks to be 2.4×10^{24}g (15),

FIG. 1 - The organic carbon content of shales (A) and of average sediments (B and C) as a function of time. 1) Average for 69 metasediments from Isua (Schidlowski, unpublished results); 2) averages for 3 shales and 81 various sediments from the Swaziland Sequence (28); 3) average for 406 Archean shales from the Canadian Shield (6); 4) average for 29 sediment samples, Hamersley Group (Hayes et al., unpublished results); 5) average for 326 Aphebian shales, Canadian Shield (6); 6) averages for 460 paragneisses and meta-argillites and 1,408 sediment samples from Proterozoic 1 - 2 of Russian Platform (30); 7) averages for 34 composites of 1,226 shales and 83 composites of 2,694 sediment samples from Proterozoic 3 of the Russian Platform (30). Phanerozoic record according to (29) and (39).

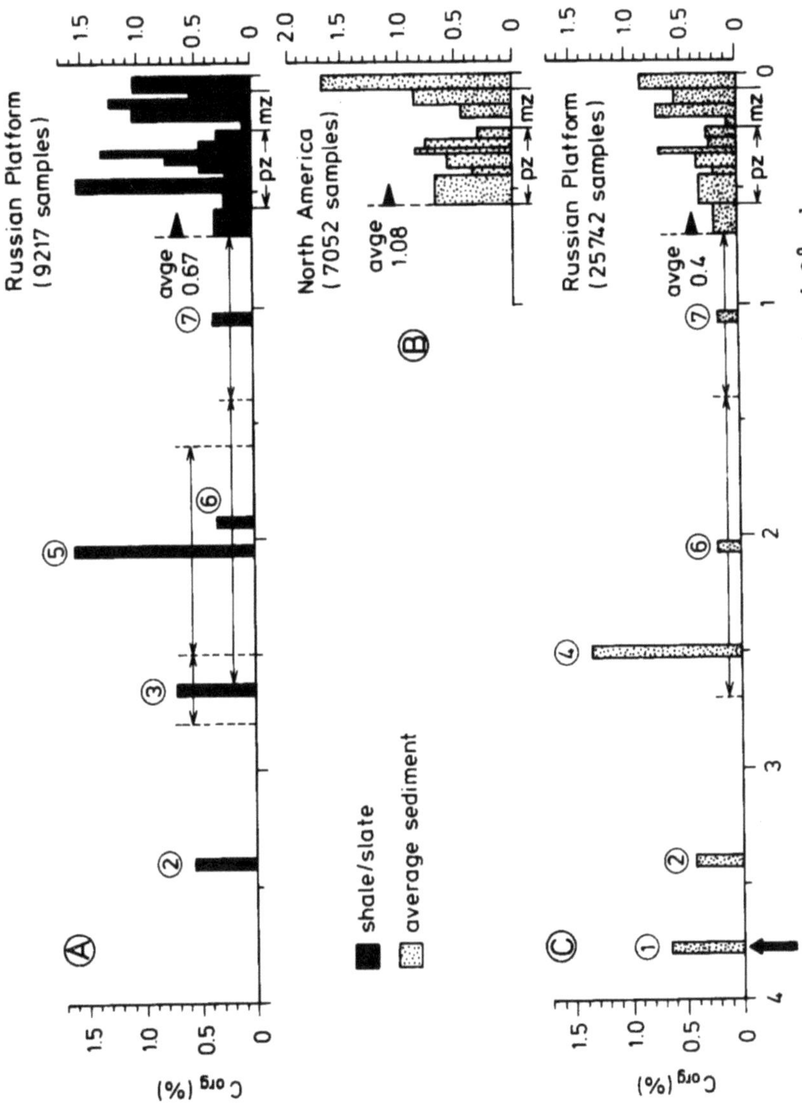
Russian Platform
(9217 samples)
avge
0.67
North America
(7052 samples)
avge
1.08
Russian Platform
(25742 samples)
avge
0.4
shale/slate
average sediment
Start of record
Age [10^9 yr]
C_{org} (%)
pz
mz
A
B
C

the C_{org} inventory of the sedimentary shell as a whole would range between 1.2 and 1.4 x 10^{22}g. It is interesting to note that the quantity of carbon that is economically recoverable is several orders of magnitude smaller. Some 10^{19}g C are recoverable in the form of coal, and between 10^{17} and 10^{18}g C as a constituent of hydrocarbons (10,18).

Previous investigators (29,39) have suggested that the organic content of Phanerozoic rocks varies considerably with time, shown by the C_{org} trend both of shales and of average sediments (Fig. 1). The tenor of C_{org} in carbonates and sandstones follows the same trend with a proportionately damped amplitude (29). Altogether, four maxima (Ordovician, Lower Carboniferous, Jurassic/Lower Cretaceous, Tertiary) and one particularly pronounced minimum (Triassic) have been reported. The similarity of the curves for the American and Russian Platforms has been interpreted to imply that these variations were world-wide; corroborating data from other parts of the world are not available.

If the above trends are real, the burial rate of C_{org} in sediments must have varied during the last 600 million years. Such assumption is corroborated by oscillations of the isotope record of marine carbonates over the same time span (41) whose positive excursions indicate periods of increased burial of organic carbon (cf. isotope mass balance, Eq. 1)). However, the two functions obviously lack the degree of covariance required to effectively support each other (Fig. 2). Accordingly, the reality of the C_{org} variations shown in Fig. 1 may still be doubted, a final solution of the problem surely calling for a considerable enlargement of the data base. In principle, moderate variations in the C_{org} content of sediments through time (usually believed to imply concomitant changes in the C_{org}/C_{carb} ratio) could be expected to result from the chemical inertia of the atmosphere-ocean-biosphere-crust system. Immediate causes might be changes in primary productivity in response to varying supplies of limiting inorganic nutrients (notably phosphorus) by the weathering cycle, fluctuations in

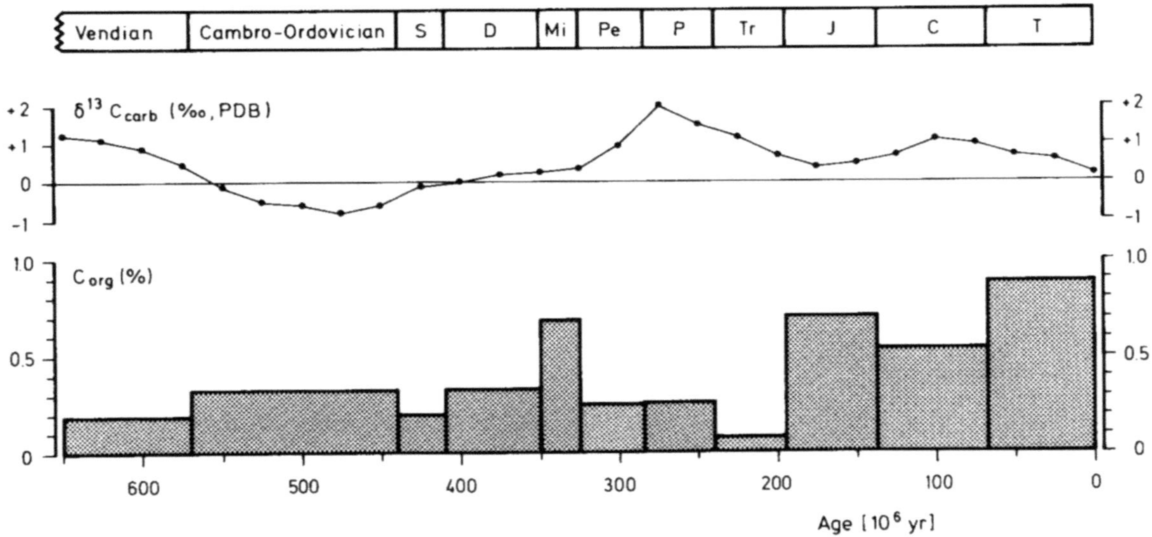

FIG 2 - Organic carbon content of Phanerozoic sediments of the Russian Platform (29) compared with the isotope record of sedimentary carbonates (36,41). According to the isotope mass balance of Eq. 1, increased burial rates of C_{org} over time spans of 10^5 yr (residence time of marine bicarbonate) would drive $\delta^{13}C_{carb}$ toward positive values; hence, maxima in C_{org} should be coupled with positive extremes in $\delta^{13}C_{carb}$. In the actually measured functions such correlation is, however, largely absent though, in a more qualitative way, the higher organic carbon content of Carboniferous and younger rocks is reflected by positive values of $\delta^{13}C_{carb}$ while the lower C_{org} averages of Early Paleozoic sediments are mostly accompanied by negative ones. Because of the rapid mixing of the marine bicarbonate reservoir, the $\delta^{13}C$ values of marine carbonates integrate over the whole exogenic exchange reservoir and hence are more likely to convey a reliable signal of the chemical state of the exogenic carbon system than do C_{org} assays from limited regions.

the fraction of the standing biomass destined to end up as sedimentary organic matter, or changes in the overall sedimentation rate through time.

Compared to the relative wealth of Phanerozoic data, the quality of the Precambrian record is deplorably poor (cf. Fig. 1). The only quantitatively important results are the C_{org} values recently obtained for some Aphebian and Archaean shales from the Canadian Shield (6) and for sediments from the Proterozoic of the Russian Platform (30). The means yielded by this relatively small number of Precambrian sediments generally lie within the range of C_{org} in Phanerozoic formations; this suggests that the average C_{org} content of these sediments does not differ greatly from that of geologically younger rocks. The scarcity of the data base for the Precambrian will, naturally, conceal secular fluctuations like those claimed for the Phanerozoic record.

THE ISOTOPIC COMPOSITION OF ORGANIC MATTER IN SEDIMENTARY ROCKS

Any low-temperature reduction of oxidized carbon in terrestrial near-surface environments is primarily due to photosynthetic fixation of CO_2 and HCO_3^-. This reduction is achieved largely by means of the Calvin (or reductive pentose phosphate, RPP) cycle in which CO_2 is reduced to the carbohydrate level. The first step in this process is the fixation of CO_2 by the enzyme ribulose-1,5-biphosphate (RuBP) carboxylase as a 3-carbon carboxylic acid (phosphoglyceric acid, PGA).

In both this enzymatic carboxylation reaction and in the preceding uptake and diffusion of CO_2 to the reaction sites, a kinetic isotope effect favors the reaction rates (and thereby preferential metabolization) of the light carbon isotope (^{12}C) (25). Fractionation in the carboxylation reaction is considerably larger than during CO_2 uptake and intracellular transport. Hence, the bias in favor of ^{12}C in photosynthesis principally derives from the properties of the enzyme RuBP carboxylase (23). Total fractionation of carbon isotopes depends on which of the two reactions is rate-controlling; it may therefore vary from plant to plant and in response to environmental factors. This is illustrated in Fig. 3 which summarizes the isotopic composition of carbon in some major groups of higher plants, algae, and autotrophic bacteria. It should be noted that the C4 and CAM pathways both appeared relatively late in angiosperm evolution (39); their impact on the carbon cycle probably dates from Late Mesozoic times. As a whole, the terrestrial biomass seems to be enriched by about 25‰ in light carbon compared to the inorganic carbon pool of the crust-ocean system mainly composed of carbonate and bicarbonate with a $\delta^{13}C$ value close to 0‰.

The isotopic composition of organic matter may change by several permil in either direction during decomposition and burial in sediments. This is due in large part to the reconstitution of the organic compounds during their maturation pathway (14). An isotopically light hydrocarbon fraction is removed preferentially with increasing thermocatalytic stress, producing an enrichment of 1 to 4‰ in the ^{13}C content of the residual kerogen (26,32). However, the $\delta^{13}C$ of diagenetically stabilized kerogen is rarely sufficiently different from that of its precursor compounds to obscure its biological pedigree. High-rank (partly graphitized) kerogens from metamorphic terranes, whose $\delta^{13}C$ values may be markedly displaced in the positive direction (12,16,22,40), are noteworthy exceptions to this rule.

The isotopic composition of carbon in kerogen preserved in the sedimentary record permits us to trace the isotopic composition of fossil organic matter into the geological past (Fig. 4). During the Phanerozoic, the standard deviations of $\delta^{13}C_{org}$ for selected age groups lie basically between -20 and - 30‰. The <u>average</u> for all Precambrian kerogens apparently lies within the same limits, but the total spread seems to be larger; negative extremes of $\delta^{13}C_{org}$ sometimes exceed -40‰. Since Precambrian kerogens are principally derived from bacteria and algae rich in isotopically light (-24 to -32‰) lipids and relatively poor in "heavy" (-15 to -22‰) carbohydrates, a

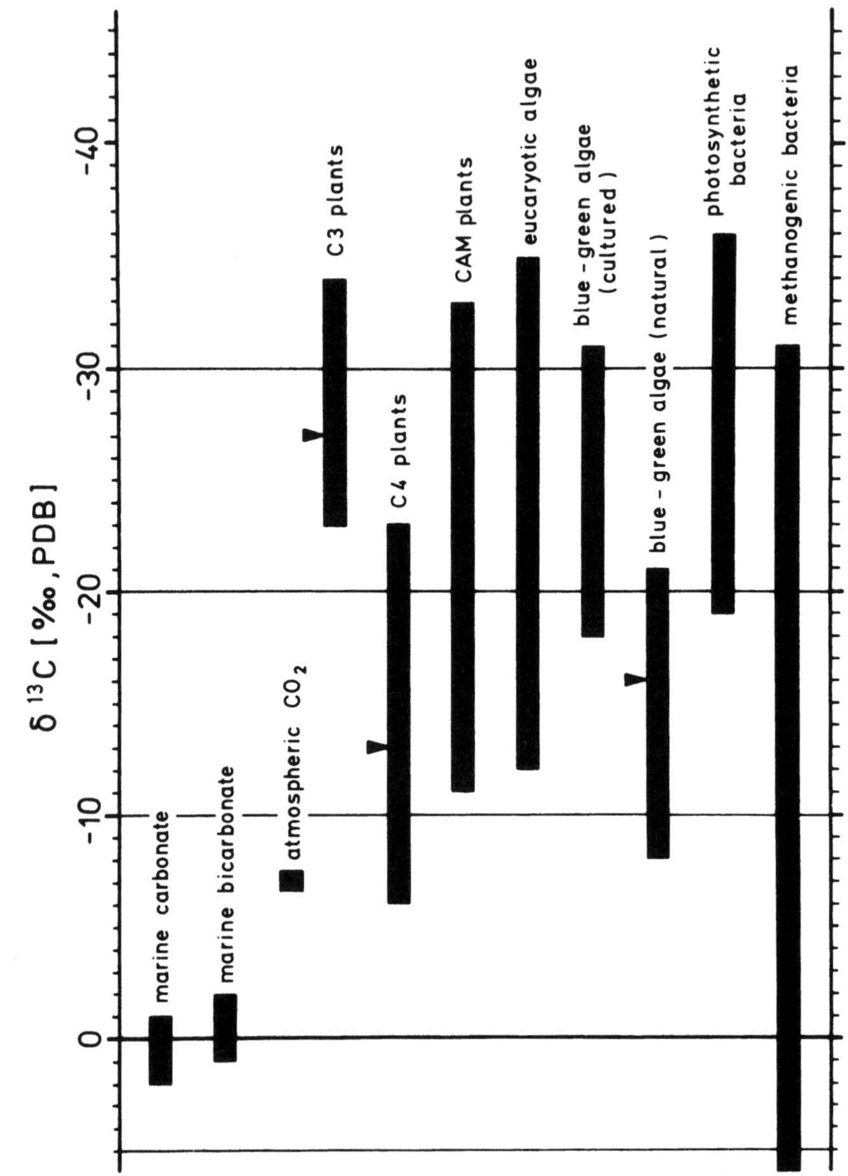
δ ¹³C [‰, PDB]
0
-10
-20
-30
-40
marine carbonate
marine bicarbonate
atmospheric CO₂
C4 plants
C3 plants
CAM plants
eucaryotic algae
blue-green algae (cultured)
blue-green algae (natural)
photosynthetic bacteria
methanogenic bacteria

FIG. 3 - The isotopic composition of higher plants, algae, and autotrophic bacteria ((3,13,23,24,45,46), and others) compared with the environmental reservoirs of atmospheric CO_2 and marine bicarbonate (HCO_3^-). Note that average plant matter is enriched in ^{12}C by about 20-30‰ compared to oceanic bicarbonate, the most abundant inorganic carbon species in the exogenic exchange reservoir (which is subsequently preserved as sedimentary carbonate with but minor isotope change). The large spread of $\delta^{13}C_{org}$ values is in marked constrast to the narrow range in $\delta^{13}C_{carb}$. Note that this difference is preserved in the sedimentary record (cf. Fig. 4).

slight negative displacement of the most ancient $\delta^{13}C_{org}$ values is rather to be expected but this cannot account for the occasional negative spikes between -35 to -50‰ . Such light values have hitherto only been reported in connection with methane, and it is almost certain that bacteria of the methane cycle were involved in the formation of these kerogens (35). Extremely light ($\delta^{13}C_{org} \approx -80$‰), non-volatile organics have also been found in Phanerozoic sediments (19). The positive extremes are most probably due to metamorphism, the chance of which is proportionately increased in progressively older rocks (22,34,40). Low H/C ratios typical of high-rank kerogens tend to support a metamorphic explanation for the unusually heavy organics (e.g., Isua (35)).

As a whole, the scatter of sedimentary $\delta^{13}C_{org}$ values in Fig. 4 basically reflects the variations displayed by major groups of extant plants and autotrophic microorganisms, thus conveying a remarkably consistent isotopic signal of biological activity as from 3.5×10^9 yr ago (12,33,35). This signal is, moreover, supported by the small variations of the marine $\delta^{13}C_{carb}$ record tethered to the zero permil line within ± 2‰ (as $\delta^{13}C_{org}$ and $\delta^{13}C_{carb}$ are coupled by an isotope mass balance (Eq. 1, see below), a basic parallelism between the two records would be a necessary corollary of this interpretation). Since the isotope shifts observed for both sedimentary carbon species in the Isua supracrustals can be accounted for by amphibolite-grade metamorphism (1), the above statement may hold for the complete time span $t \lesssim 3.8 \times 10^9$ yr (34). The uniformity of the scatter band back to $\sim 3.5 \times 10^9$ yr ago would, furthermore,

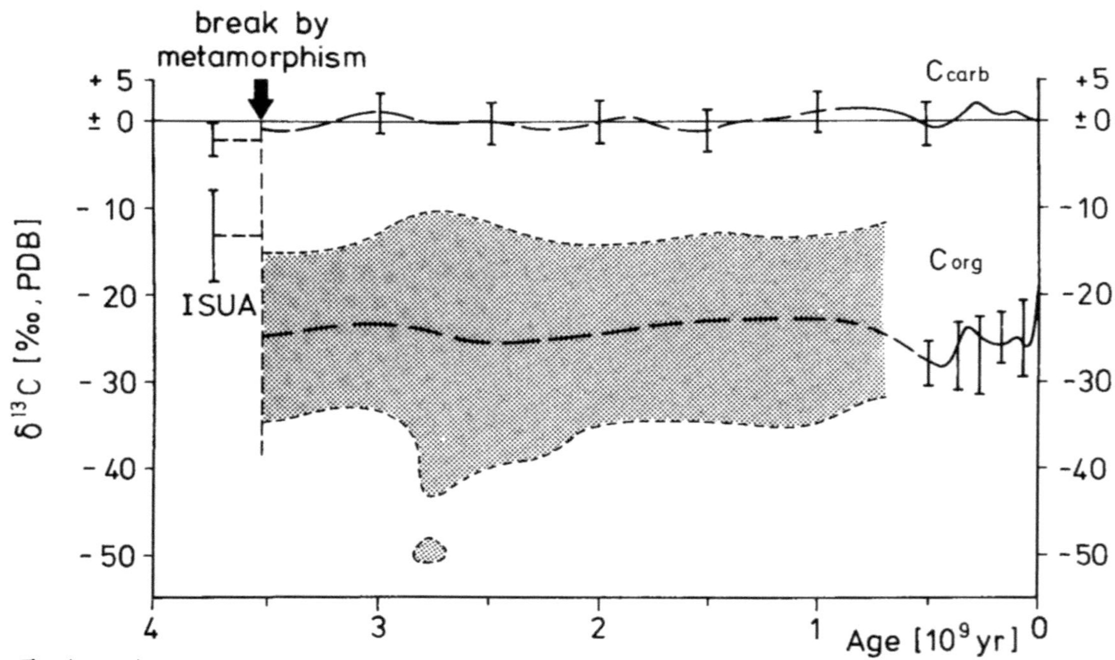

Fig. 4 - The isotopic composition of organic carbon (C_{org}) and carbonate carbon (C_{carb}) over geologic time ((8,12, 14,34,35,41,43) and others). Dashed lines for Precambrian functions indicate a lower confidence level compared to that for the Phanerozoic ($t < 0.6 \times 10^9$ yr) record. Note difference between small standard deviation for $\delta^{13}C_{carb}$ ($+0.5 \pm 2.6$ ‰) and large scatter for $\delta^{13}C_{org}$ (about -26 ± 6 ‰) which reflects similar differences in the precursor materials (see Fig. 3). Phanerozoic age functions for carbonates (41) and organic carbon (14,43) vary sympathetically with different amplitudes; bars superimposed on $\delta^{13}C_{org}$ curve are standard deviations for selected time intervals from a different source (8). The Precambrian $\delta^{13}C_{org}$ record is represented as a total spread, the scatter field (shaded area) summarizing data from (35). The extremely light kerogens from the 2.7×10^9 yr old Fortescue Group (Hamersley Basin, Australia) have counterparts in the Phanerozoic falling even outside the δ-scale (-80 to -90 ‰, cf. (19)).

give testimony to an extreme conservatism of the basic biochemical mechanisms of carbon fixation, notably the Calvin cycle (12,35).

The only challenge to this interpretation notably of the unmetamorphosed sedimentary $\delta^{13}C_{org}$ record would be the operation of an inorganic geochemical process able to mimic biological fractionations with a remarkable degree of precision. However, such process must have also equalled biological carbon fixation in its quantitative capacity in order to bring the carbonate values of the exogenic exchange reservoir up to zero permil from an initial value of about -5‰ in accordance with the isotope mass balance (Eq. 1). It is considered unlikely that an inorganic process could meet these requirements. Although fractionations in Fischer-Tropsch-type processes go in the same direction as photosynthesis, their magnitude (up to 100‰) considerably exceeds that observed in the sedimentary record (20). Only fractionations reported for Miller-Urey-type spark discharge syntheses (mainly between 4 and 12‰, cf. Fig. 5) marginally overlap the range exhibited by sedimentary $\delta^{13}C_{org}$ values (7). The virtual absence of substantially reduced average fractionations in the unmetamorphosed record would, however, exclude abiotic production of organic matter as a major kerogen source during the last 3.5×10^9 yr. With the isotope shifts displayed by both C_{org} and C_{carb} in the Isua sediments (cf. Fig. 4) consistent with a metamorphic overprint, we are even extremely hesitant to propose that such a mechanism could have worked during Isua times.

As the Isua carbonates indicate the presence of abundant CO_2 in the atmosphere 3.8×10^9 yr ago, CO- and CH_4-based abiogenic production of organic matter is unlikely to have played a significant role ever since then. Moreover, reasonable doubt may be raised as to whether spark discharge syntheses could have produced organic substances in quantities comparable to those accrued as a result of biological activity. Preliminary C_{org} assays (Fig. 1C, No. 1) as well as the occurrence of abundant

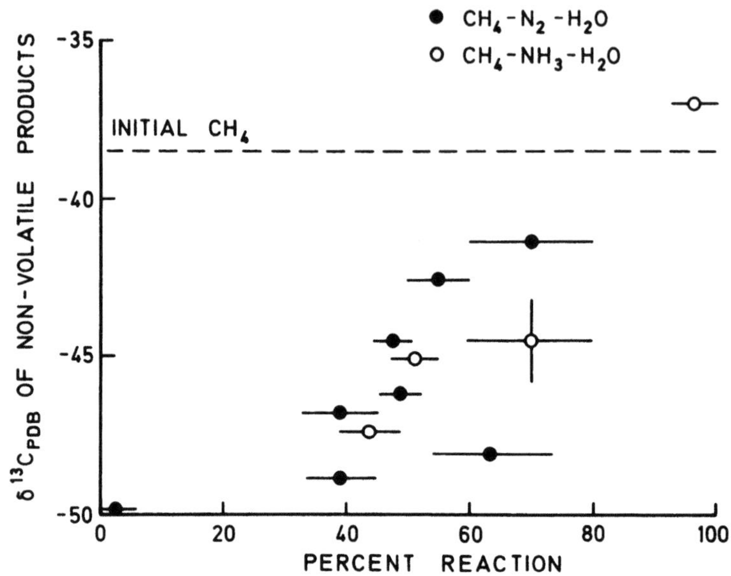

FIG. 5 - Carbon isotope fractionation in Miller-Urey spark discharge processes (7). Reaction products examined consist of all material that was non-volatile at 10^{-3}mm and 300°K. This non-volatile fraction may be analogous to protokerogenous material in prebiotic sediments. Note that the fractionation observed at low reactant consumption just overlaps the range attributed to biological fractionations in the sedimentary record.

graphitic seams (notably in the banded iron formation) indicate that the organic carbon content of the Isua supracrustals does not differ significantly from that of geologically younger rocks. Hence, unless hitherto unidentified detrital carbon constituents from prebiological times are hidden in the bulk organics, the possibility that an isotopic record of abiogenic synthesis is preserved in the Isua rocks must be rated as rather tenuous.

The morphological record of life confirms the inferences from carbon isotopes in sedimentary rocks. The record of cellular structures and stromatolites goes back to about 3.5×10^9 yr ((2,9,21,42), see also Awramik, this volume). On the other hand, microstructures of suspected biological affinities recently described from the Isua suite (26) are objects of a current controversy (5). The high level of diversification displayed by the 3.5×10^9 yr old Warrawoona microfossils (2) and the "modern" carbon isotope geochemistry inferred for the pre-metamorphic Isua rocks make it almost certain, however, that life originated a considerable length of time prior to the onset of the generally accepted fossil record.

CARBON ISOTOPE MASS BALANCE

The ratio of organic carbon to carbonate carbon in sediments is coupled to the isotopic composition of both carbon species by the ^{13}C mass balance equation

$$\delta^{13}C_p = R\delta^{13}C_{org} + (1-R)\delta^{13}C_{carb} \quad (1)$$

Here, $\delta^{13}C_p$ denotes the isotopic composition of the primordial carbon input to the system (generally assumed to be close to -5 ‰), and R is the ratio $C_{org}/(C_{org} + C_{carb})$ in sediments.

With the $\delta^{13}C$ means of organic and carbonate carbon always close to -25‰ and 0‰ over the last 3.5 x 10^9 yr (Fig. 4), the fractionation ΔC between the two carbon species stored in the sedimentary shell is largely identical with the average fractionation inherent in photosynthesis, i.e., $\Delta C = \delta^{13}C_{carb} - \delta^{13}C_{org} \cong 25‰$. Solving Eq. 1 for R would yield R = 0.2 which value implies a partitioning of total sedimentary carbon between C_{org} and C_{carb} in the ratio 0.2:0.8 or 1:4 respectively. Direct determinations of this ratio, notably in the sedimentary cover of the Russian Platform (30,31), have come close to this value.

Since the isotopic fractionation between organic and carbonate carbon is primarily established in the exogenic exchange reservoir before being propagated into the rock section of the cycle, the relatively small variations of $\delta^{13}C_{org}$ and $\delta^{13}C_{carb}$ through time might suggest, inter alia, that the input into the exchange reservoir always stayed close to the primordial value of -5‰. However, most of the CO_2 added to the atmosphere-ocean system during the Earth's later history certainly constituted recycled crustal carbon whose isotopic composition depends on the proportion of C_{org} and C_{carb} in the recycled flux.

As recycling processes integrate over the global chemical system, they are apt to always mobilize an average crustal carbon sample which, however, is not necessarily identical in its isotopic composition to primordial carbon. Accordingly, we cannot exclude that the exogenic carbon system has adjusted, over the ages, to changes in the isotopic composition of the carbon input, maintaining the relatively constant $\delta^{13}C$ levels of marine bicarbonate and of organic carbon by adaptions in other parts of the system.

CURRENT FRONTIERS

The establishment of criteria to differentiate between "genuine" kerogen (by definition biologically-derived) and abiogenically produced kerogen-like substances is pivotal to a further elucidation of the carbon cycle in the Archean. As a first step, a systematic comparative characterization of kerogens of undoubted biological origin and the polymerized carbon constituents of carbonaceous chondrites seems to be called for; an investigation of the trace amounts of reduced carbon in igneous rocks (notably those of assumed mantle provenance) would also be desirable. Isotopic information ought to be augmented by elemental (C/H/N) and organic analysis (for details see recommendations of Awramik et al., this volume).

Another problem concerns the decomposition and possible recycling of kerogen constituents under conditions of an anoxygenic weathering cycle. Questions posed by the isotopically light kerogens (cf. Fig. 4) and by the proposed secular variations of the organic carbon content of Phanerozoic sediments (29,39) also deserve further inquiry.

Acknowledgements. This paper has benefited from the stimulating atmosphere of the Precambrian Paleobiology Research Group, University of California, Los Angeles. Support from the Deutsche Forschungsgemeinschaft (SFB 73) as well as from NASA Grant NSG 7489 and the NSF Watermann Foundation Award to J.W. Schopf is gratefully acknowledged. I am particularly indebted to S. Chang and co-workers for permission to refer to some of their recent results prior to publication.

REFERENCES

(1) Allaart, J.H. 1976. The pre-3760 Myr old supracrustal rocks of the Isua area, central West Greenland, and the associated occurrence of quartz-banded ironstone. In The Early History of the Earth, ed. B.F. Windley, pp. 177-189. London: Wiley.

(2) Awramik, S.M.; Schopf, J.W.; Walter, M.R.; and Buick, R. 1981. Filamentous fossil bacteria 3.5 Ga-old from the Archaean of Western Australia. Science: in press.

(3) Benedict, C.R. 1978. The fractionation of stable carbon isotopes in photosynthesis. What's New in Plant Physiology 9: 13-16.

(4) Bolin, B.; Degens, E.T.; Kempe, S.; and Ketner, R. 1979. The Global Carbon Cycle. New York: Wiley.

(5) Bridgwater, D.; Allaart, J.H.; Schopf, J.W.; Klein, C.; Walter, M.R.; Barghoorn, E.S.; Strother, P.; Knoll, A.H.; and Gorman, B.E. 1981. Microfossil-like objects from the Archaean of Greenland: A cautionary note. Nature 289: 51-53.

(6) Cameron, E.M., and Garrels, R.M. 1980. Geochemical compositions of some Precambrian shales from the Canadian Shield. Chem. Geol. 28: 181-197.

(7) Chang, S.; Des Marais, D.; Mack, R.; Miller, S.L.; and Strathearn, G. 1981. Prebiotic organic synthesis and the origin of life. In Origin and Evolution of Earth's Earliest Biosphere, ed. J.W. Schopf. Princeton, NJ: Princeton University Press, in press.

(8) Degens, E.T. 1969. Biogeochemistry of stable carbon isotopes. In Organic Geochemistry, eds. G. Eglinton and M.T. Murphy, pp. 304-329. Berlin: Springer.

(9) Dunlop, J.S.R.; Muir, M.D.; Milne, V.A.; and Groves, D.I. 1978. A new microfossil assemblage from the Archaean of Western Australia. Nature 274: 676-678.

(10) Durand, B. 1980. Kerogen. Paris: Editions Technip.

(11) Eglinton, G., and Calvin, M. 1967. Chemical fossils. Sci. Am. 216: 32-43.

(12) Eichmann, R., and Schidlowski, M. 1975. Isotopic fractionation between coexisting organic carbon-carbonate pairs in Precambrian sediments. Geochim. Cosmochim. Acta 39: 585-595.

(13) Fuchs, G.; Thauer, R.; Ziegler, H.; and Stichler, W. 1979. Carbon isotope fractionation by Methanobacterium thermoautotrophicum. Arch. Microbiol. 120: 135-139.

(14) Galimov, E.M. 1980. $^{13}C/^{12}C$ in kerogen. In Kerogen, ed. B. Durand, pp. 271-299. Paris: Editions Technip.

(15) Garrels, R.M., and Lerman, A. 1977. The exogenic cycle: reservoirs, fluxes and problems. In Global Chemical Cycles and Their Alteration by Man, ed. W. Stumm, pp. 23-31. Berlin: Abakon.

(16) Hoefs, J., and Frey, M. 1976. The isotopic composition of carbonaceous matter in a metamorphic profile from the Swiss Alps. Geochim. Cosmochim. Acta 40: 945-951.

(17) Holland, H.D. 1978. The Chemistry of the Atmosphere and Oceans. New York: Wiley.

(18) Hunt, J.M. 1972. Distribution of carbon in crust of Earth. Bull. Am. Assoc. Petrol. Geol. 56: 2273-2277.

(19) Kaplan, I.R., and Nissenbaum, A. 1966. Anomalous carbon isotope ratios in nonvolatile organic material. Science 153: 744-745.

(20) Lancet, M.S., and Anders, E. 1970. Carbon isotope fractionation in the Fischer-Tropsch synthesis and in meteorites. Science 170: 980-982.

(21) Lowe, D.R. 1980. Stromatolites 3.400-Myr old from the Archaean of Western Australia. Nature 284: 441-443.

(22) McKirdy, D.M., and Powell, T.G. 1974. Metamorphic alteration of carbon isotopic composition in ancient sedimentary organic matter: new evidence from Australia and South Africa. Geology 2: 591-595.

(23) O'Leary, M.H. 1981. Carbon isotope fractionation in plants. Phytochemistry: in press.

(24) Pardue, J.W.; Scalan, R.S.; Van Baalen, C.; and Parker, P.L. 1976. Maximum carbon isotope fractionation in photosynthesis by blue-grren algae and a green alga. Geochim. Cosmochim. Acta 40: 309-312.

(25) Park, R., and Epstein, S. 1960. Carbon isotope fractionation during photosynthesis. Geochim. Cosmochim. Acta 21: 110-126.

(26) Peters, K.E.; Rohrback, B.G.; and Kaplan, I.R. 1980. Laboratory-simulated thermal maturation of recent sediments. In Advances in Organic Geochemistry 1979, eds. A.G. Douglas and J.R. Maxwell, pp. 547-557. Oxford: Pergamon.

(27) Pflug, H.D. 1978. Yeast-like microfossils detected in oldest sediments of the Earth. Naturwissenschaften 65: 611-615.

(29) Reimer, T.O.; Barghoorn, E.S.; and Margulis, L. 1979. Primary productivity in an Early Archaean microbial ecosystem. Precambrian Res. 9: 93-104.

(30) Ronov, A.B. 1958. Organic carbon in sedimentary rocks (in relation to the presence of petroleum). Geochemistry 1958: 510-536.

(31) Ronov, A.B. 1980. Osadotchnaya obolotchka zemli (20th Vernadski Lecture). Moscow: Izdatel'stvo Nauka.

(31) Ronov, A.B., and Yaroshevski, A.A. 1969. Chemical composition of the Earth's crust. Am. Geophys. Union Geophys. Mon. Ser. 13: 37-57.

(32) Sackett, W.M.; Nakaparksin, S.; and Dalrymple, D. 1968. Carbon isotope effects in methane production by thermal cracking. In Advances in Organic Geochemistry 1966, eds. G.D. Hobson and G.C. Speers, pp. 37-53. Oxford: Pergamon.

(33) Schidlowski, M. 1980. Antiquity of photosynthesis: possible constraints from Archaean carbon isotope record. In Biogeochemistry of Ancient and Modern Environments, eds. P.A. Trudinger and M.R. Walter, pp. 47-54. Berlin: Springer.

(34) Schidlowski, M.; Appel, P.W.U.; Eichmann, R.; and Junge, C.E. 1979. Carbon isotope geochemistry of the 3.7 x 10^9 yr old Isua sediments, West Greenland: implications for the Archaean carbon and oxygen cycles. Geochim. Cosmochim. Acta 43: 189-199.

(35) Schidlowski, M.; Hayes, J.M.; and Kaplan, J.R. 1981. Isotopic inferences of ancient biochemistries. In Origin and Evolution of Earth's Earliest Biosphere, ed. J.W. Schopf. Princeton, NJ: Princeton University Press, in press.

(36) Schidlowski, M., and Junge, C.E. 1981. Coupling among the terrestrial sulfur, carbon and oxygen cycles: numerical modeling based on revised Phanerozoic carbon isotope record. Geochim Cosmochim. Acta 45: 589-594.

(37) Smith, B.N. 1976. Evolution of C4 photosynthesis in response to changes in carbon and oxygen concentrations in the atmosphere through time. BioSystems 8: 24-32.

(38) Tissot, B.P., and Welte, D.H. 1978. Petroleum Formation and Occurrence. Berlin: Springer.

(39) Trask, P.D., and Patnode, H.W. 1942. Source Beds of Petroleum. Tulsa: American Association of Petrol. Geol.

(40) Valley, J.W., and O'Neil, J.R. 1981. $^{13}C/^{12}C$ exchange between calcite and graphite: a possible thermometer in Grenville marbles. Geochim. Cosmochim. Acta 45: 411-419.

(41) Veizer, J.; Holser, W.T.; and Wilgus, C.K. 1980. Correlation of $^{13}C/^{12}C$ and $^{34}S/^{32}S$ secular variations. Geochim. Cosmochim. Acta 44: 579-587.

(42) Walter, M.R.; Buick, R.; and Dunlop, J.S.R. 1980. Stromatolites 3.400-3.500 Myr old from the North Pole area, Western Australia. Nature 284: 443-445.

(43) Welte, D.H.; Kalkreuth, W.; and Hoefs, J. 1975. Age-trend in carbon isotopic composition in Paleozoic sediments. Naturwissenschaften 62: 482-483.

(44) Whittaker, R.H., and Likens, G.E. 1973. Carbon in the biota. In Carbon and the Biosphere, eds. C.M. Woodwell and E.V. Pecan, pp. 281-302. Washington, D.C.: U.S. Atomic Energy Commission.

(45) Wong, W.W., and Sackett, W.M. 1978. Fractionation of stable carbon isotopes by marine phytoplankton. Geochim. Cosmochim. Acta 42: 1809-1815.

(46) Wong, W.W.; Sackett, W.M.; and Benedict, C.R. 1975. Isotope fractionation in photosynthetic bacteria during carbon dioxide assimilation. Plant Physiol. 55: 475-479.

(47) Woodwell, G.M.; Whittaker, R.H.; Reiners, W.A.; Likens, G.E.; Delwiche, C.C.; and Botkin, D.D. 1978. The biota and the world carbon budget. Science 199: 141-146.

Mineral Deposits and the Evolution of the Biosphere, eds. H.D. Holland and M. Schidlowski, pp. 123-154. Dahlem Konferenzen, 1982.
Berlin, Heidelberg, New York: Springer-Verlag.

The Composition of Kerogen and Hydrocarbons in Precambrian Rocks

D. M. McKirdy* and J. H. Hahn**
*South Australian Department of Mines and Energy
P. O. Box 151, Eastwood, S. A. 5063, Australia
**Max-Planck-Institut für Chemie, 6500 Mainz, F. R. Germany

Abstract. Prior to 1975 geochemical investigations of organic matter in Precambrian rocks were concerned primarily with two questions: Is it biogenic? If so, is it also syngenetic? Only now, with the application of modern concepts of petroleum genesis to the study of hydrocarbons and kerogen in a variety of Precambrian sedimentary and low-grade metamorphic rocks, does the development of satisfactory criteria for biogenicity and syngeneity appear feasible. Recent advances include the recognition of four major kerogen types, documentation of the manner in which their chemical and carbon isotopic composition changes during catagenesis and anchimetamorphism, and the finding of probable remnants of algal and/or bacterial lipids, carbohydrates, and proteins among the pyrolysis and ozonolysis products of kerogens as old as 2.8 x 10^9 yr (2.8 Gyr). n-Alkane distributions in Precambrian shales, carbonates, cherts, and banded ironstones are strongly controlled by the lithofacies and rank of the host rock. Acyclic isoprenoid alkanes are the most widely reported and definitive "biological marker" hydrocarbons. The oldest known steranes and triterpanes would appear to be those in Riphean crude oils from Siberia. Despite the common occurrence of organic matter in Precambrian metallic ore deposits, a detailed knowledge of its composition is still lacking.

INTRODUCTION

Bona fide microfossils and stromatolites have been found in sediments as old as 3.4-3.5 Gyr (2,43,62). Thus, biogenic organic matter (preserved mainly as kerogen and associated hydrocarbons) can be expected to occur widely in Archaean and Proterozoic sedimentary rocks, and the alleged inorganic sources of hydrocarbons and graphitic carbon in certain Precambrian metamorphic and intrusive igneous rocks (6,35,44,49) are called into question. It is conceivable that by the time the 3.76 Gyr-old Isua metasediments in Greenland were deposited, the sedimentary organic carbon reservoir was already about 50-60% as large as at present (56). The main precursors of this early Precambrian organic matter were probably primitive autotrophs.

A basic tenet of the organic geochemical studies of Precambrian rocks that were undertaken during the 1960s and early 1970s (23,35,44) was that chemical fossils (13), i.e., organic compounds which have survived diagenesis and subsequent burial with their molecular structure (or isotopic composition) substantially unchanged from the time they were part of a living organism, could provide valuable information on ancient life processes and the major events in the early evolution of the biosphere. Because they are easily isolated and analyzed, hydrocarbons (in particular alkanes) were the most commonly studied components of Precambrian organic matter (35). However, it was soon realized that hydrocarbons and other solvent-extractable compounds, especially when present in trace amounts, need not have the same age as their host rock (58). The bulk of the dispersed organic matter in ancient rocks (kerogen) is particulate, insoluble, relatively inert, and hence more likely to be both syngenetic and free of post-lithification contamination. These same attributes make the analysis of kerogen difficult except by techniques that are rather indirect, non-selective, or severe: white light and fluorescence-mode microscopy; determination of elemental composition and stable isotope ratios (e.g., of C, H, and N); infrared spectroscopy (IR); ^{13}C-nuclear magnetic resonance (NMR); and gas chromatography-mass spectrometry (GC-MS) of the products of pyrolysis, oxidation, and hydrogenation.

This paper examines the progress made since 1974 in determining the composition and origin of kerogen and hydrocarbons in Precambrian rocks. The relevant earlier work was reviewed by McKirdy (35).

KEROGEN

During the sedimentation and early diagenesis of organic matter, biochemical macromolecules (carbohydrates, proteins, lipids, and pigments) are partially or wholly broken down by inorganic processes (e.g., hydrolysis) and bacterial enzymes into their component biomonomers (sugar, amino acids, fatty acids, etc.). Some of these simple compounds condense to form a range of lipidic and humic complexes. These random geopolymers, together with intact remnants of biopolymers, are the precursors of kerogen (28,59). The formation and preservation of protokerogen is favored where the residence time of organic detritus in the overlying (preferably oxygen-depleted) water column is short, and where acid or neutral, anoxic conditions prevail at and below the sediment-water interface (10,43). The present composition of a kerogen is a function of the nature of its principal biochemical precursors (whether hydrogen-rich lipids or oxygen-rich carbohydrates and proteins), its depositional environment (Eh, pH), and its integrated time-temperature history. The well established relationship between burial temperature, duration of heating, and rank (degree of coalification) (59) implies that syngenetic Precambrian kerogens are likely to be extensively altered unless burial has never been deep, and the kerogen has not been exposed to temperatures much in excess of 60-70°C.

Kerogen Types and Their Maturation

The composition of Precambrian (and early Palaeozoic) kerogens is surprisingly diverse. At least four types can be recognized (39). Type I kerogen is hydrogen-rich; on a van Krevelan diagram (Fig. 1) its composition plots along the coalification track of alginite. Analysis by pyrolysis-hydrogenation-gas chromatography (PHGC) gives a pyrogram in which n-alkanes up to C_{25} (or higher, depending on kerogen rank and the experimental conditions employed) are dominant; simple aromatic hydrocarbons (benzene, toluene, ethylbenzene, and the xylenes) occur in very low concentrations relative to aliphatic moieties of equivalent carbon number. Likely precursors of this highly aliphatic organic matter are lipid-rich planktonic chlorophytes (e.g., Tasmanites, Gloecapsomorpha), recognized petrologically as discrete

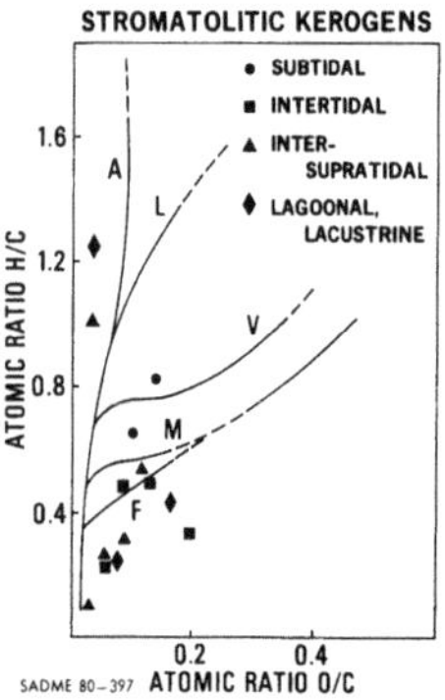

FIG. 1 - Van Krevelen diagram showing the elemental composition of Precambrian and Palaeozoic stromatolitic kerogens (38,39). Lines are coalification tracks of coal macerals: A = alginite (Type I kerogen); L = liptinite (Type II kerogen); V = vitrinite, M = macrinite (Type III kerogen); F = fusinite (Type IV kerogen).

algal bodies (alginite A) and thin films of lamellar alginite (alginite B) derived from bloom or mat-forming cyanobacteria, partially reworked by anaerobic bacteria and therefore incorporating lipid-rich bacterial cells (38,39). Type I kerogens have been reported from several Cambrian and older rocks for which anoxic depositional environments may be inferred: the evaporitic, alkaline playa-lacustrine facies of the Observatory Hill Beds, Officer Basin, South Australia (atomic H/C = 1.37-1.19) (38); and marine phosphorites of the Beetle Creek Formation, Georgina Basin, Queensland (atomic H/C = 1.13), the Lesser Karatau deposit, USSR (atomic H/C = 0.53), and the Areyonga Formation, Amadeus Basin, Northern Territory (atomic H/C = 0.53) (39,51). The low H/C value of the latter kerogens reflects their deeper burial and consequent higher degree of thermal alteration. Another Precambrian example of highly altered Type I kerogen (alginite B) may be the granules and seams (up to 5 cm thick) of thucholite associated with placer gold, uraninite, and pyrite in conglomerates and quartz arenites of the ≃ 2.7 Gyr-old Witwatersrand Group (42). The thucholite has an atomic H/C value ≃ 0.56 (7); on pyrolysis in vacuo it yields mainly alkylbenzenes, alkylnaphthalenes, and aromatic sulfur compounds (44,66). This unusual material has an exceptionally high content of organic free radicals; this is consistent with a severe modification of its original (?aliphatic) structure by prolonged exposure to ionizing radiation emanating from inclusions of detrital and/or chemically precipitated uraninite.

Petrologically, Type II microbial kerogens comprise bituminite (amorphous sapropelized algal and protozoan remains), lipto-detrinite (fragmented planktonic algal cells and acritarchs), or a mixture of lipid-rich (Type I) and humic (Type III: see below) organic matter. On a van Krevelen plot they occupy an intermediate position which overlaps the liptinite coalification track, as shown (Fig. 2) by two kerogens from fossiliferous black shales of the 1.45 Gyr-old McMinn Formation, McArthur Basin, Northern Territory (48). Kerogen from the famous 1.1

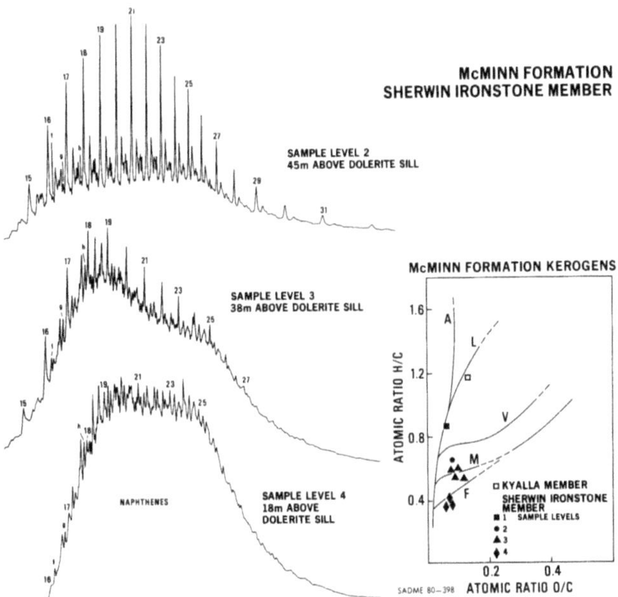

FIG. 2 - Kerogen composition and selected alkane patterns in shales of the McMinn Formation (37,48). Samples from the Sherwin Ironstone Member illustrate the effects of accelerated maturation by a 50m thick dolerite sill. Key: f = norpristane (C_{18}); g = pristane (C_{19}); h = phytane (C_{20}); numbers refer to carbon number of n-alkanes.

Gyr-old Nonesuch Shale, Michigan, likewise is probably Type II organic matter (4). PHGC analysis of the Kyalla and Sherwin Ironstone (level 1) kerogen shows that it differs little in aromaticity from Type I kerogen but that it is considerably more naphthenic; n-alkyl chains are still prominent, but they are shorter and are present in lower concentration relative to branched and cyclic alkyl moieties. The straight-chain aliphatic character of the McMinn (Kyalla) kerogen is also evident from the C_{11}-C_{21} n-alkanoic acids obtained by oxidation with chromic acid; no branched acids were detected (23). This contrasts with the Nonesuch kerogen, which yielded a complex mixture of branched and normal acids after similar oxidative treatment, and a high proportion of isoalkanes on pyrolysis. Sporopollenin, a chemically resistant biopolymer in the cell walls of certain algae, is a possible precursor of Type II kerogen, although its importance as a source of Precambrian organic matter has been overstated (7). The maturation of Type II kerogen is illustrated by samples from the Sherwin Ironstone Member at four stratigraphic levels above an intrusive dolerite sill (Fig. 2). Its composition changes, possibly via a series of coalification jumps (39), from that of liptinite at level 1 (55 m above sill), through micrinite at levels 2 and 3, to fusinite at level 4 (18 m above sill). IR and PHGC data indicate a progressive shortening of aliphatic chains and a loss of alkyl groups, accompanied by aromatization of alicyclic rings and increased condensation of the kerogen carbon skeleton. Type II kerogen is "oil-prone" (as is the less common Type I variety), and is likely to have been the principal source of indigenous Precambrian petroleum (see below). It has been isolated from bituminous shales and carbonates of the 0.8 Gyr-old Bitter Springs Formation, Amadeus Basin (atomic H/C = 0.85-0.80) (37), and from several units within the $\simeq$1.5 Gyr-old McArthur Group, McArthur Basin (atomic H/C = 0.92-0.75) ((55), and J.D. Saxby, unpublished results). The latter kerogens occur in sediments that are host to important Pb-Zn sulfide mineralization of both the fine-grained stratiform variety (e.g., the H.Y.C. deposit) and the coarse-grained stratabound Mississippi Valley type (e.g., Coxco), and commonly contain bitumen in vugs and fractures (61).

Oxygen-rich, hydrogen-poor kerogen is preserved in very low concentrations in certain Proterozoic to mid-Palaeozoic stromatolitic carbonates and cherts (Fig. 1) (36). Such organic matter (designated Type III or Type IV kerogen) is most probably derived from cyanobacterial mucilage and from algal and bacterial pectic tissue (39). Primary Type III kerogen, exemplified by the organic matter in some subtidal stromatolites, has an elemental composition similar to that of vitrinite, whereas Type IV kerogen is more analogous to fusinite and presumably underwent partial oxidation during early diagenesis. It is important to distinguish primary humic kerogens from the aromatic coalification residues of lipid-rich algal organic matter which, as shown in Figure 2, may mimic the composition of either micrinite (secondary Type III) or fusinite (secondary Type IV). Phytoclasts of the latter kerogen may have a microcoke texture typical of rank fusinite. Type IV kerogens contain small amounts of volatile carbon that is anomalously aliphatic for organic matter so deficient in hydrogen (Fig. 3). The aliphatic component of these and other humic Precambrian kerogens commonly displays an odd carbon-number predominance in the C_{10+} range which may be a biological marker for bacterial cell-wall lipids (39).

Biological Markers in Kerogen Degradation Products

Vacuum pyrolysis-GC-MS, a technique which minimizes inter- and intramolecular rearrangements of pyrolysis products, is particularly suitable for the identification of specific biological marker

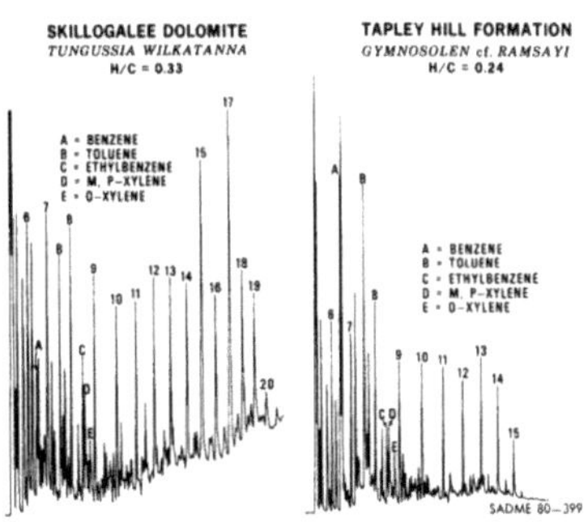

FIG. 3 - PHGC traces of Precambrian stromatolitic kerogens (39).

compounds in kerogen pyrolysates. Its use in the analysis of kerogens in the Bulawayan (≈2.8 Gyr), Rupemba-Belingwe (≈2.7 Gyr), and Transvaal (≈2.3 Gyr) stromatolites from southern Africa (44,57) has yielded, among other compounds (mainly alkylbenzenes and alkylnaphthalenes): furan, 2-methylfuran, 2,5-dimethlyfuran, propylfuran, and furaldehyde; pyrrole, benzonitrile, and aliphatic nitriles; and C_9-C_{20} n-alkanes. These products were interpreted respectively as remnants of biological carbohydrates, proteins, and fatty acids. The presence of 2-n-propyl-3-methyltetrahydrofuran and 2-n-propyltetrahydropyran among the products of the ozonolysis of the Transvaal kerogen is further evidence that carbohydrates were indeed major precursors of Precambrian stromatolitic kerogens. The discovery of such biological markers in selected kerogens as old as 2.8 Gyr is at odds with the findings of Leventhal and others (34) who obtained only C_1-C_8 hydrocarbons from the stepwise pyrolysis of solvent-extracted samples of 30 organic-rich Precambrian sediments and concluded that "even the least metamorphosed of these ancient sediments have evolved toward amorphous carbon or graphite and do not yield useful 'biochemical fossils'." This conclusion, while unduly pessimistic in relation to the Precambrian sedimentary record as a whole, does appear to be justified for the well studied Archaean Swaziland Sequence, South Africa.

Swaziland Sequence

Organic matter is relatively abundant in these sediments (total organic carbon = 0.2-2.1%) (53), but it has undergone considerable thermal alteration. Kerogens in carbonaceous cherts of the ≈3.5 Gyr-old Onverwacht Group have elemental compositions (C d.a.f. = 86-99%, atomic H/C = 0.24-0.03) (37) which are in accord with the greenschist facies regional metamorphism of this sequence. Spuriously high H/C values (0.4-1.4) reported for certain Onverwacht kerogens (12,40) are attributable to hydrated fluorosilicates, which formed as artifacts of the kerogen isolation procedure (37). The Theespruit Formation (lower Onverwacht Group) contains graphitic kerogen which is of consistently higher rank than that in the Hooggenoeg and Kromberg Formations (upper Onverwacht Group) (Fig. 4). It is not clear whether this is due to deeper burial of the unit or to its proximity (at least where sampled) to an intrusive granite pluton. The high rank is relevant to the interpretation of the isotopic composition of carbon in these ancient kerogens (Fig. 5) (see below).

Ozonolysis and pyrolysis studies (12,23,44) of various Swaziland kerogens, mainly from cherts of the Onverwacht and Fig Tree (≃3.4? Gyr) Groups, indicate the presence of highly condensed aromatic structures incorporating short aliphatic bridges and some longer, peripheral, normal and branched alkyl chains. Fig Tree and upper Onverwacht (Kromberg, Swartkoppie) kerogens tend to be somewhat more aliphatic than those lower in the sequence. The apparent survival of long hydrocarbon chains (up to C_{15}) in the highly altered Theespruit kerogen (12) is unexpected and problematical.

Stable Carbon Isotope Ratios

An extensively documented, but still poorly understood aspect of Precambrian kerogen is the isotopic composition of its carbon.

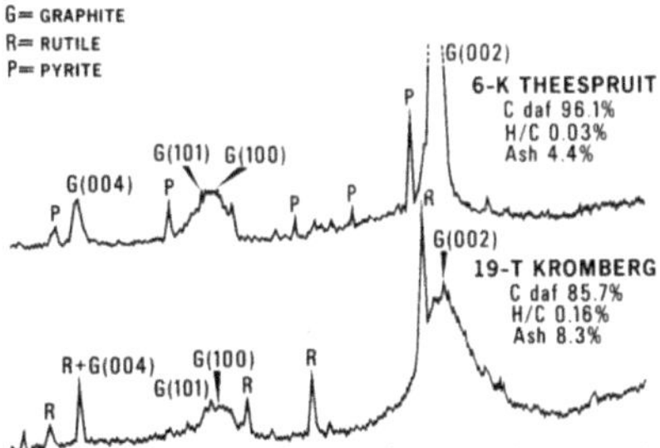

FIG. 4 - X-ray diffraction traces of kerogens from two Onverwacht Group cherts illustrating the progressive development of an incipient graphite lattice (37). Note: d.a.f. = dry, ash-free.

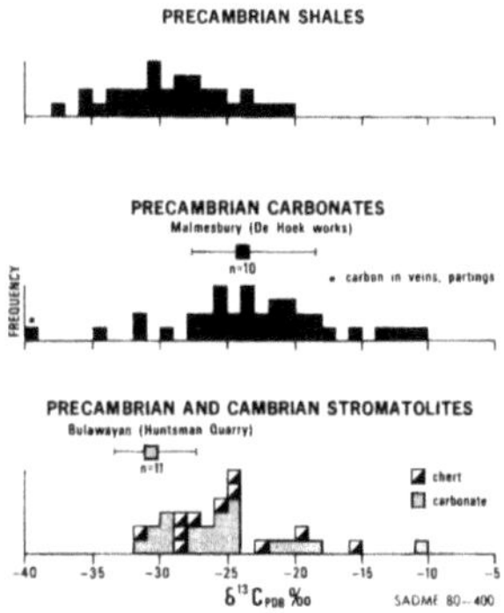

FIG. 5 - The isotopic composition of carbon in kerogen from Precambrian shales and carbonates, and from Precambrian and Cambrian stromatolites (3,14,23,29, 37,38,40,45,57). Horizontal bars give range and mean of values at the two named localities.

$\delta^{13}C$ values (expressed as per mil difference from the PDB standard) in the range of -6 to -53‰ have been reported for the total organic carbon (usually > 95% kerogen) in Precambrian sedimentary and metasedimentary rocks (9,15,20,45). Interpretation of these data is difficult, because the isotopic composition of carbon in kerogen is influenced by a complex array of factors, including the nature and growth habitat of the source biota, and the effects of thermal alteration during catagenesis and anchimetamorphism. In early Archaean rocks (e.g., Isua Supracrustals), the situation is further complicated by the possibility of abiogenic contributions to the kerogen. Nevertheless, by reference to specific suites of samples (Figs. 5-9), certain inferences can be drawn regarding the probable causes and significance of the wide range of the isotopic composition of carbon in Precambrian kerogen.

Biological fixation of CO_2 by photoautotrophs and chemoautotrophs is the major production mechanism of organic matter that is depleted in ^{13}C relative to its inorganic carbon source. The degree of fractionation ($\Delta\delta^{13}C = \delta^{13}C$ cellular carbon - $\delta^{13}C$ inorganic substrate) measured in controlled cultures of cyanobacteria and chlorophytes: -24 to -1‰ (47) is of particular relevance for understanding the origin of Precambrian kerogens. The maximum fractionation, which was achieved under conditions of high CO_2 availability (i.e., slow growth rates and/or low population density), corresponds closely to the mean fractionation between Precambrian (and Phanerozoic) organic and carbonate carbon ($\Delta\delta^{13}C \simeq 25$‰) (15,56). A higher partial pressure of CO_2 in the atmosphere may explain why Precambrian kerogen is on the whole isotopically lighter than recent paralic microbial mats (3,36). Cases of extreme ^{12}C enrichment ($\delta^{13}C$ <-28‰) may be the result of biological recycling of $^{12}CO_2$ from the anoxic decomposition of photoautotrophic remains (31); alternatively, they may be indicative of organic matter derived from methanogenic and/or methanotrophic bacteria (J.M. Hayes, personal communication).

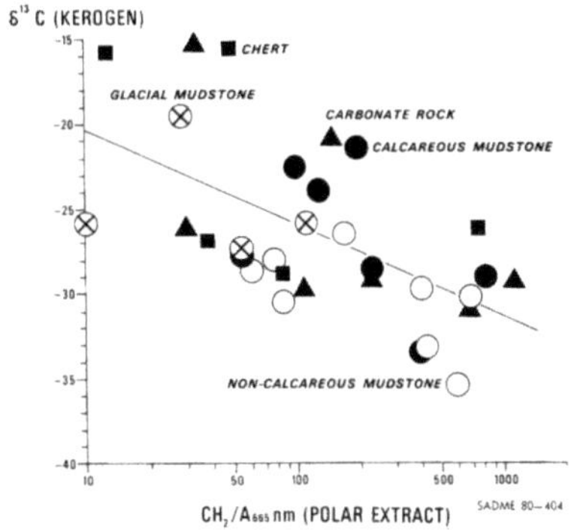

FIG. 6 - Relationship between the isotopic composition of carbon in kerogen and the "aliphaticity" of polar extract (29).

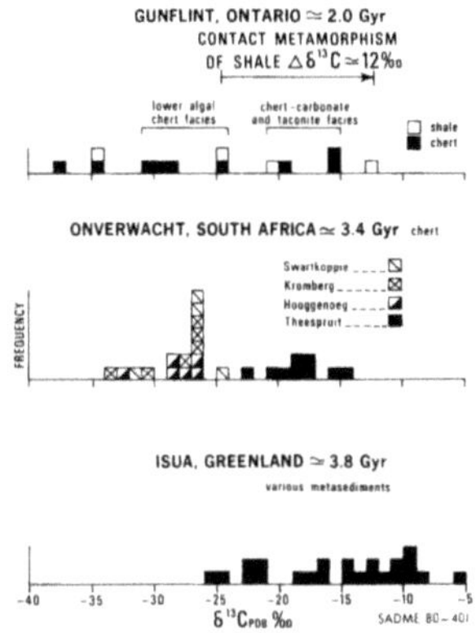

FIG. 7 - Isotopic composition of carbon in kerogen in Precambrian rocks of the Gunflint Iron Formation, Onverwacht Group, and Isua Suprastructal Belt (3,29,40, 45,46,49,56).

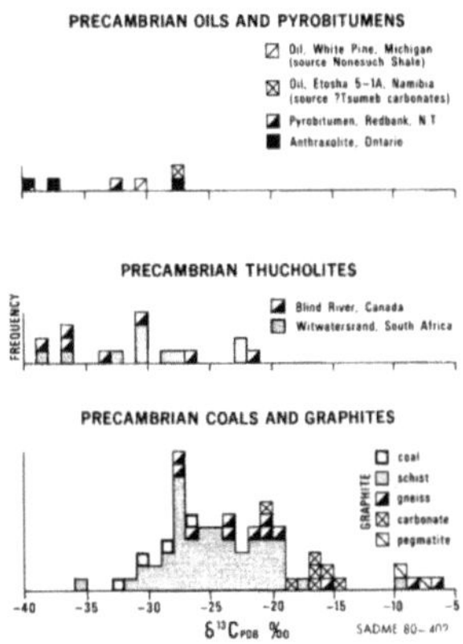

FIG. 8 - Isotopic composition of carbon in Precambrian coals, graphites, thucholites, oils, and pyrobitumens ((3,4,5,8, 14,17,22,23,32); D.M. McKirdy and P.J.M Ypma, unpublished result).

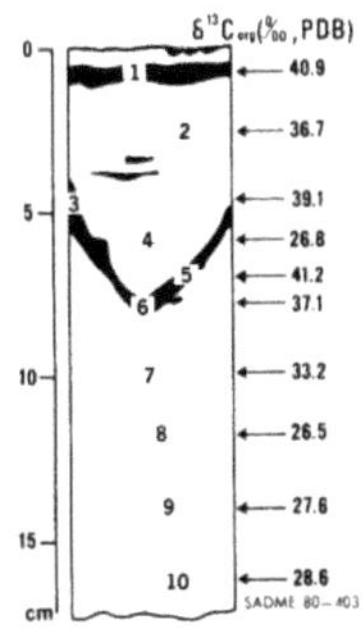

FIG. 9 - Isotopic composition of carbon in kerogen and "visible carbon" in a core of the Transvaal Dolomite (15).

Anomalously light organic carbon is preserved in rocks of the Gunflint Iron Formation of Ontario, the Ventersdorp Group of

South Africa, and the Fortescue Group of Western Australia; these units are all of similar age (2.5-2.7 Gyr) and have a similar tectonic setting (M.R. Walter, personal communication). $\delta^{13}C$ values for kerogen from the Fortescue Group range from -35 to -53‰ (20). In the Biwabik Iron Formation (a correlative of the Gunflint) coexisting reduced carbon ($\delta^{13}C$ = -31 to -35‰) and iron carbonates ($\delta^{13}C$ = -3 to -19‰) are both highly depleted in ^{13}C (50). One possible reason is the operation of post-diagenetic redox reactions of the type:

$$6\ Fe_2O_3 + {}^{12}C\ org \rightarrow 4\ Fe_3O_4 + {}^{12}CO_2\ ,$$

with subsequent exchange between $^{12}CO_2$ and adjacent carbonate minerals. A corollary of this hypothesis is that the reduced carbon now present in the iron formation (average organic carbon content = 0.2%) is a remnant of a primary kerogen which was even more depleted in ^{13}C and that was originally present in somewhat higher concentration.

There is good evidence of a link between the isotopic composition of carbon in kerogen and the nature of the source material. The existence of several Precambrian kerogen types, ranging from lipidic to humic, implies that there has been selective preservation of cellular carbon. The various biochemical fractions of an organism are isotopically different (9). Thus, lipid-based kerogens (Type I and II) are likely to be more depleted in ^{13}C ($\delta^{13}C \simeq$ -20 to -35‰) than kerogens (primary Types III and IV) derived principally from carbohydrates and proteins ($\delta^{13}C \simeq$ -15 to -25‰). This may in part explain the apparent isotopic difference between carbon in unmetamorphosed Precambrian shales and carbonates (Fig. 5). Kerogens in the former tend to be isotopically lighter, and isotopically lighter Precambrian kerogens yield more highly aliphatic polar extracts (Fig. 6) (29).

Variations in the isotopic composition of carbon in kerogen that are related to differences in their source and environment have been reported for the Gunflint Iron Formation. $\delta^{13}C$ values for the shallow water algal chert facies of the Gunflint are significantly different (Fig. 7) from those of the deeper water chert-carbonate and taconite facies; the differences can be correlated with differences in the original microbiotas (3).

Post-depositional maturation can give rise to solid organic matter which is isotopically either lighter or heavier than the parent kerogen. This fact renders the 30‰ spread in $\delta^{13}C$ values for Precambrian graphites (Fig. 8) a little less problematical. Accumulations of carbonaceous matter which represent mobilized portions of a dispersed, low-rank kerogen are enriched in ^{12}C because of the preferential rupture of ^{12}C - ^{12}C bonds during thermal cracking and the loss of ^{13}C-bearing carboxyl groups from resins and asphaltenes during subsequent devolatilization and condensation (40). A chromatographic effect which discriminates against ^{13}C compounds may also be involved (15). Anthraxolite from Ontario, other pyrobitumens, and possibly certain thucholites (Fig. 8), and veins and partings of "carbon" in the Transvaal Dolomite (Fig. 9) are examples of such organic matter from Precambrian rock units. The $\delta^{13}C$ value of the residual, immobile kerogen changes little with advancing maturation until a rank of atomic H/C $\approx$ 0.30 ($\cong$86% C d.a.f.) is attained (Fig. 10). At higher rank, however, there is a positive shift which becomes most marked in kerogens with H/C values <0.15 ($\cong$91% C d.a.f.). Notable examples of such isotopically heavy graphitic kerogen occur in the lower Onverwacht Group, and in the amphibolite-grade Isua metasediments (Fig. 7). Metamorphism is capable of increasing kerogen $\delta^{13}C$ values by at least 12‰ (Fig. 7). For this metamorphic adjustment of carbon isotopic composition to proceed via cracking of $^{12}CH_4$ from hydrogen-deficient organic matter (40), an external source of hydrogen (e.g., H_2O) is required. Preferential oxidation of kerogen ^{12}C to CO_2 in the presence of water has been proposed as one possible mechanism (21). The apparent lack of metamorphic alteration of $\delta^{13}C$ in some high-rank kerogens (Fig. 10) may be indicative of the non-availability of an external hydrogen source. Possibly, these are cases of "dry" incipient metamorphism.

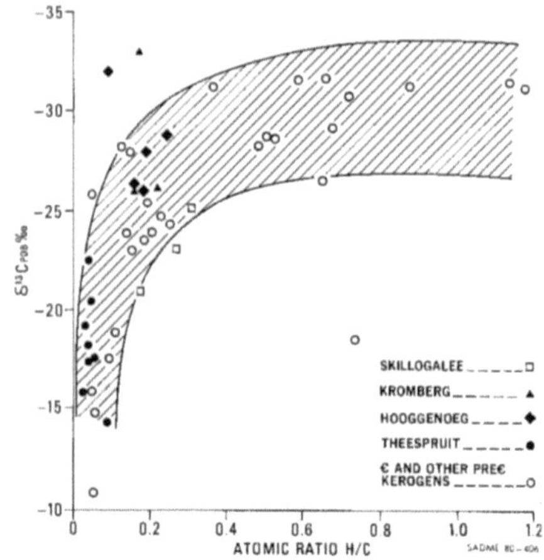

FIG. 10 - Relationship between the H/C atomic ratio and carbon isotopic composition of kerogen in cherts of the Onverwacht Group, South Africa, and Skillogalee Dolomite, South Australia. Shaded trend based on data from Cambrian and other Precambrian rocks (37).

The broad range (up to 15‰) of $\delta^{13}C$ values for samples from the same formation and locality (Fig. 5), and thus presumably of the same rank, clearly demonstrates that valid inferences regarding the source, environment, or thermal history of organic carbon in a particular geologic unit cannot be made on the basis of a single kerogen $\delta^{13}C$ determination.

HYDROCARBONS

Indigenous saturated and aromatic hydrocarbons occur in rocks of all ages, from Recent to Archaean. Unsaturated aliphatic hydrocarbons (alkenes) are more labile; when they are present in Precambrian rocks, they are likely to be of relatively recent origin.

Biogenic Versus Abiogenic Hydrocarbons

Although rare occurrences of abiogenic bitumen (including hydrocarbons) may exist in certain igneous rocks (35), the evidence overwhelmingly favors a biogenic origin for the C_{2+} hydrocarbons preserved in most Precambrian sedimentary and metamorphic rocks.

Biological lipids (various hydrocarbons, fatty acids, and alcohols) and pigments (chlorophyll, carotenoids) are the main primary source of C_{15+} sedimentary hydrocarbons. During diagenesis, biolipids undergo chemical reactions, including hydrolysis, reduction and decarboxylation, and attack by bacterial lipolytic enzymes (28). The resulting geolipids consist of hydrocarbons and other compounds with fewer double bonds and functional groups than their biolipid precursors. However, in Precambrian and other ancient rocks, the bulk of the extractable hydrocarbons (which include aromatics and low molecular weight alkanes not found in living organisms or recent sediments) are products of the catagenetic alteration of the non-hydrocarbon geolipids and associated kerogen.

Since carbon-carbon bonds are fairly resistant to thermal cracking (particularly in the absence of clay catalysts), hydrocarbons have considerable potential as chemical or molecular fossils (13). Biosynthetic lipids and pigments possess characteristic carbon skeletons (Fig. 11) which are preserved as the corresponding saturated or aromatic hydrocarbons under favorable

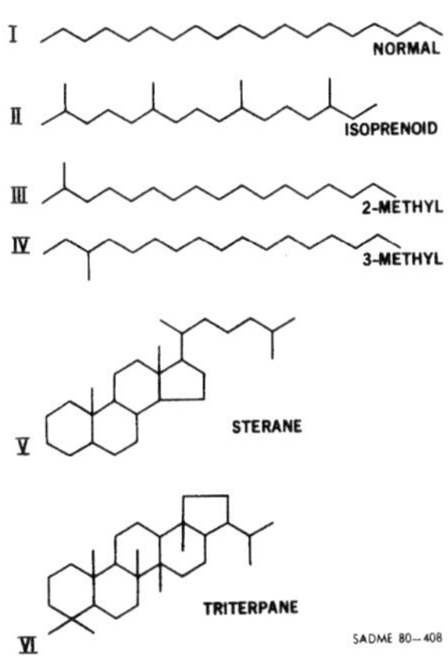

FIG. 11 - Carbon skeletons of various types of hydrocarbon chemical fossils found in ancient sediments.

geological circumstances, and may thus serve as biological markers (30).

Recognition of such compounds in Precambrian rocks presupposes knowledge of potential hydrocarbon precursors in extant examples of those microorganisms which first appeared during Precambrian time (35,36). An adequate appreciation of the effects of diagenetic and catagenetic processes on hydrocarbon structure, stereochemistry, and molecular-weight distribution is equally important.

Catagenesis of Hydrocarbons

Numerous studies of petroleum formation (28,59) have documented the manner in which hydrocarbons are generated, modified, and destroyed during catagenesis. Thermally immature, fine-grained Precambrian sedimentary rocks will contain only low concentrations of hydrocarbons; these consist of C_{15+} hydrocarbons (≤10-15 mg/g C) that were inherited more or less directly from biolipid precursors and of small amounts of C_2-C_7 hydrocarbons. The ratio of pristane to n-heptadecane (pr/n-C_{17}) and of phytane to n-octadecane (ph/n-C_{18}) may be relatively high (≥1), and the concentration of phytane commonly exceeds that of pristane (pr/ph<1) (37). Such rocks are rare, however, and are confined to tectonically undisturbed parts of Late Precambrian sedimentary basins.

Most Precambrian organic-rich sediments have crossed the threshold of intense hydrocarbon generation; some still lie within the "oil window." Examples of the latter, some of which are associated with oil shows or accumulations, occur in the Lena-Tunguska province of the Siberian Platform (41); the Nonesuch Shale (4,28); the McArthur and Amadeus Basins (37,48); the Etosha Basin, Namibia (65); and the sub-salt carbonate sequence of the Arabian Shield (M.R. Walter and J.H. Oehler, personal communication). These sediments contain a complete range of hydrocarbons typical of unbiodegraded crude oil. Of the C_{15}-C_{20} acyclic isoprenoids, pristane (C_{19}) is preferentially generated during early catagenesis, although pr/ph values rarely exceed 2 in mature (and postmature) Precambrian sediments (37). The contrasting C_{15+} alkane distributions of oil-mature shale (kerogen H/C = 0.85) and carbonate (kerogen H/C = 0.73) from the Bitter Springs Formation (Fig. 12) demonstrate the influence of host lithofacies on hydrocarbon composition. A similar contrast is evident in the C_{12+} alkanes of shale-derived (Fig. 13) and carbonate-derived (Fig. 14) Precambrian crude oils, although the unusual Tsumeb oil also shows signs of immaturity (prominent naphthene hump in sterane-triterpane region) and/or partial biodegradation (relative lack of n-alkanes).

Extensive disproportionation of hydrocarbons occurs during catagenesis and anchimetamorphism (52). Clays in argillaceous rocks become more acidic and facilitate catalytic cracking of alkyl chains by a carbonium ion mechanism. Branched and cyclic alkanes are the preferred C_{15+} products (Fig. 2). The overall yield of C_{15+} saturates plus aromatics decreases in both shales and carbonates as hydrocarbons of lower molecular weight are generated and expelled.

Hydrocarbons as Precambrian Biological Markers

The thermal stability and appreciable concentration of alkanes in non-reservoir rocks (up to several hundred ppm) and crude oils (>60% of C_{12+} fraction) makes them arguably the best available molecular fossils. Their diagnostic value as biological markers varies inversely with their structural and stereochemical complexity as follows: pentacyclic triterpanes ≈ steranes > acyclic isoprenoids > iso, anteiso alkanes > n-alkanes (Fig. 11). A biological marker may be a discrete compound or a homologous series of compounds (Table 1).

Steranes found in the 2.7 Gyr-old Soudan Shale (30) do not appear to be syngenetic (23). Despite reports of tetracyclic alkanes in the C_{15+} extract of the Nonesuch Shale (28) and steranes in the Nonesuch oil (Fig. 13) (4), subsequent careful study of the oil (24,26) has revealed no trace of steranes or triterpanes.

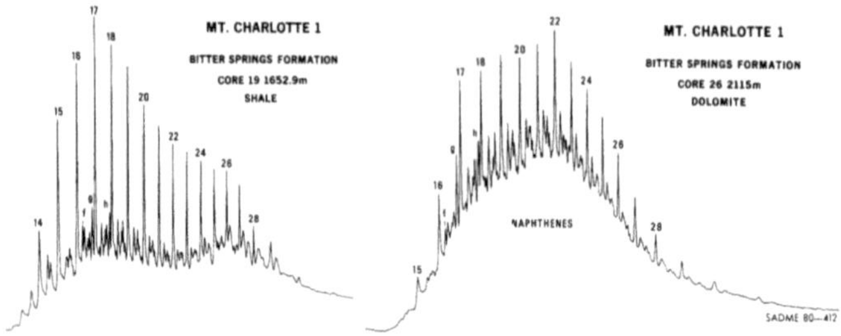

FIG. 12 - Gas chromatograms of the C_{15+} total alkanes in extracts of shale and carbonate from the Bitter Springs Formation, Amadeus Basin, Northern Territory (37).

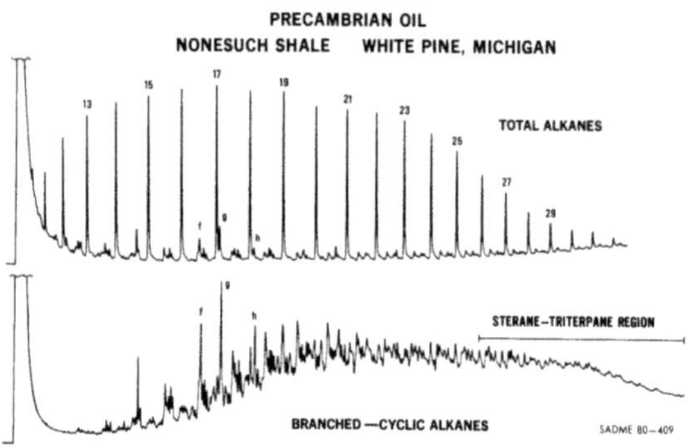

FIG. 13 - Gas chromatograms of C_{12+} total and branched/cyclic alkane fractions of the Nonesuch Shale oil seep, Michigan, USA (24).

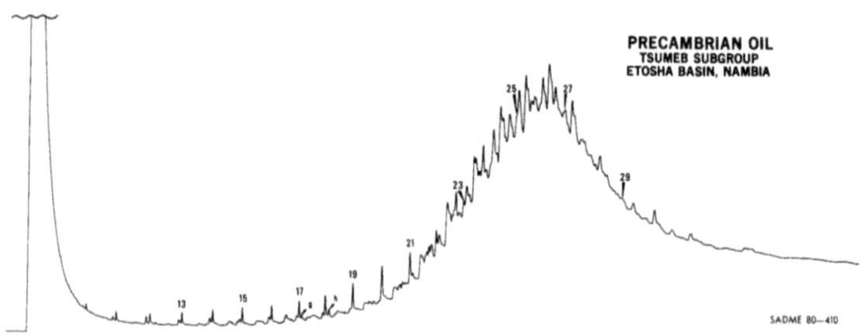

FIG. 14 - Gas chromatogram of the C_{12+} total alkanes in oil from the Tsumeb Subgroup, Etosha Basin, Namibia (D.M. McKirdy and P.J.M. Ypma, unpublished result).

TABLE 1 - Hydrocarbons with potential as Precambrian biological markers.

Hydrocarbon type	Characteristic feature	Possible source(s)
normal alkanes	C_{12}-C_{21} with odd carbon-number preference C_{15} or C_{17} dominant	algae, bacteria, cyanobacteria
	C_{22}-C_{35} with odd carbon-number preference	chlorophytes of Botryococcus type, certain cyanobacteria
	C_{20}-C_{32} with even carbon-number preference	anaerobic bacteria
	C_{22} prominent	dinoflagellates, zooplankton (via diagenetic reduction of $C_{22:6}$ fatty acid)
branched alkanes	7- and 8-methylheptadecane	cyanobacteria
	C_{16}-C_{30} iso and anteiso alkanes	bacteria
	C_{13}-C_{20} regular isoprenoids	photosynthetic algae, certain bacteria
	C_{21}-C_{40} regular and irregular isoprenoids (incl. squalane)	archaebacteria
cyclic alkanes	C_{27}-C_{35} pentacyclic triterpanes of hopane series	prokaryotes
	steranes	algae, cyanobacteria
	4-methylsteranes	dinoflagellates, methanotrophic bacteria
	n-alkylcyclohexanes	algae (via diagenetic intramolecular cyclization of n-fatty acids), thermoacidophilic bacteria
aromatic	methyl-branched alkylbenzenes	archaebacteria
	mono- and polyaromatic steranes and triterpanes	algae, bacteria

If they were originally present, they may have undergone dehydrogenation to related mono- or polyaromatic hydrocarbons.

Hollerbach and Welte (26) found low concentrations of steranes (<30 ppm) but no triterpanes in 3 Cambrian oils, whereas Early

Cambrian, non-marine carbonates and associated oil from the Officer Basin (Fig. 15) contain both steranes and a series of C_{27}-C_{35} hopanoid triterpanes (38). The ability of syngenetic steranes and triterpanes to survive in rocks at least as old as 0.68 Gyr is demonstrated by their presence in crude oils from the Riphean of the Siberian Platform (1). The unusual sterane distribution of these ancient oils (viz. C_{29} stereoisomers markedly dominant) may reflect their derivation from cyanobacteria. Limited data (11,54) suggest that sitosterols (C_{29}) commonly are the major sterol components in extant cyanobacteria.

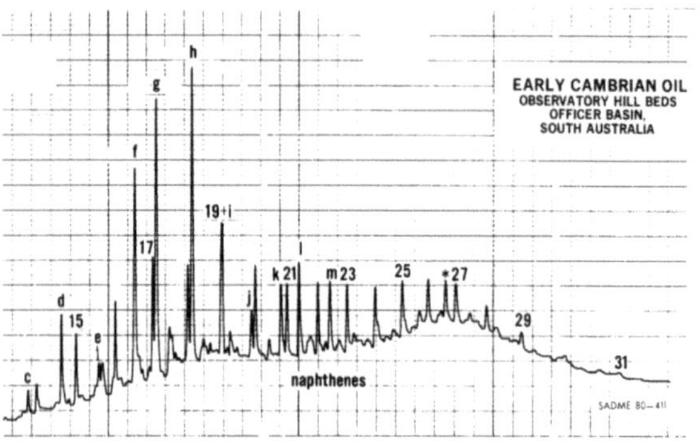

FIG. 15 - Gas chromatogram of the C_{15+} total alkanes in oil from the Observatory Hill Beds, Officer Basin, South Australia (38). Key: c - m = C_{15} - C_{25} acyclic regular isoprenoids. Asterisked peak = squalane.

Acyclic C_{14}-C_{20} regular isoprenoids, notably pristane and phytane, have been identified in most of the Precambrian sedimentary rocks so far examined (35). Because their most likely biochemical precursor is the phytyl (C_{20}) alcohol side-chain of chlorophyll, they are commonly regarded as evidence of photosynthesis. However, several alternative sources may have been important in extreme (i.e., strictly anoxic, hypersaline, or acid hot springs) Precambrian sedimentary environments. Pristane and lesser amounts of phytane are present as discrete hydrocarbons in certain non-photosynthetic bacteria (36). Phytanyl (C_{20}) diethers occur in the polar lipids of halophilic and methanogenic archaebacteria (60). In sediments to which archaebacteria have contributed organic matter, the higher acyclic isoprenoids (C_{21+}) are also likely to be present: methanogens and thermoacidophiles contain dibiphytanyl (C_{40}) ethers and C_{21}-C_{30} isoprenoid alkenes (and minor alkanes), whereas the irregular $C_{30:6}$ isoprenoid squalene and related hydrosqualenes are major neutral lipids in both halophiles and methanogens (27,60). The high concentration of isoprenoid alkanes in the Officer Basin oil (Fig. 15) reflects its derivation from halophilic and/or methanogenic archaebacteria which subsisted on the remains of mat-forming cyanobacteria (38). A similar archaebacterial source appears likely for the isoprenoids recovered from fluid inclusions in quartz crystals within Namibian Precambrian metasediments (33). The C_{14}-C_{20} isoprenoids were identified, but the published alkane chromatogram also shows the probable presence of C_{21}-C_{25} sesterterpanes and squalane. The C_{21} isoprenoid occurs in the Nonesuch oil (30). Among other hydrocarbon molecular fossils in this oil, the iso and anteiso alkanes (C_{16}-C_{28}) suggest but do not prove a derivation from bacterial lipid precursors, and the n-alkylcyclohexanes (C_{16}-C_{23}) may be diagenetic products of algal long-chain unsaturated fatty acids (24,30).

To date, aromatic hydrocarbons in Precambrian rocks have been all but ignored by organic geochemists, despite their potential as biological markers (Table 1). The occurrence of apparently penecontemporaneous 3:4-benzpyrene (among other polycyclic aromatics) in seams of amphibole asbestos within banded iron formations (BIF's) of the 2.3 Gyr-old Transvaal Supergroup (19) is interesting, although its alleged microbial origin remains uncertain.

n-Alkane Patterns in Selected Precambrian Rocks

The structural simplicity of n-alkanes detracts from their utility as biological markers. Nevertheless, certain features of their C_{12+} distribution may be indicative of a biosynthetic origin: the position of the maximum (or maxima) and the predominance of odd or even carbon-numbered homologues (see below). A modest but significant recent advance in Precambrian organic geochemistry is the discovery that sedimentary n-alkane patterns are strongly dependent on lithofacies (Figs. 12, 16-19) (16,37, 64). This finding is important for two reasons. First, it opens the way to more detailed studies of Precambrian palaeoenvironments using other, possibly more sensitive, marker compounds. Second, the recognition of post-lithification contamination (23,35) becomes much easier if the distribution of Precambrian alkanes varies in a predictable way with rock type and degree of catagenesis.

The use of vaporization-gas chromatography (VGC) (18) instead of solvent-extraction procedures is also a distinct advance. This technique has been improved by the introduction of glass capillary columns and the addition of a mass spectrometer, and is now capable of analyzing volatile organic matter (VOM) up to C_{30} in rock samples of 100 gm and less. By avoiding solvents, the chances of sample contamination in the laboratory are reduced, as are evaporative losses of C_{12}-C_{20} compounds. n-Alkane profiles of VOM in samples from the 2.2 Gyr-old Malmani Dolomite, Transvaal, are shown in Figs. 16 and 17. The formation consists of 1500m of dolomite interbedded with stromatolitic and oolitic limestone and thin bands of black shale. The n-alkane distribution of upper intertidal facies dolomite is rather narrow, with a maximum at C_{17}, and similar to those of various dolomites from the lower intertidal facies (Level 10, Fig. 16). n-Heptadecane, the major hydrocarbon in many Precambrian rocks, is usually attributed to cyanobacterial precursors (35,36). The n-alkane profiles of dolomites from the lower subtidal facies (Level 11, Fig. 16) are strikingly different, in that they display a marked predominance of C_{16}, C_{18}, and C_{20}. Such a distribution pattern may indicate the migration of hydrocarbons from the black shale bands in the sequence, since their n-alkanes likewise generally exhibit a strong even carbon-number predominance. This unusual feature is consistent with deposition under strictly anoxic conditions resulting in the chemical reduction of algal and/or cyanobacterial C_{16}-C_{20} straight-chain fatty acids to the corresponding even carbon-numbered n-alkanes (63).

Brooks (6) reported a distinct even carbon-numbered preference in the C_{18}-C_{26} n-alkanes from a suite of Archaean shales and ultrabasics. At Tweefontein, Western Transvaal, all the facies of the Malmani Dolomite are characterized by bimodal n-alkane patterns with maxima at C_{15}-C_{17} and C_{22}-C_{24} (Fig. 17). Similar bimodal patterns have been found in other Precambrian rocks (6,35). Their exact genetic significance is not yet clear, although a dual photosynthetic algal (including cyanobacterial)/ anaerobic bacterial source is possible.

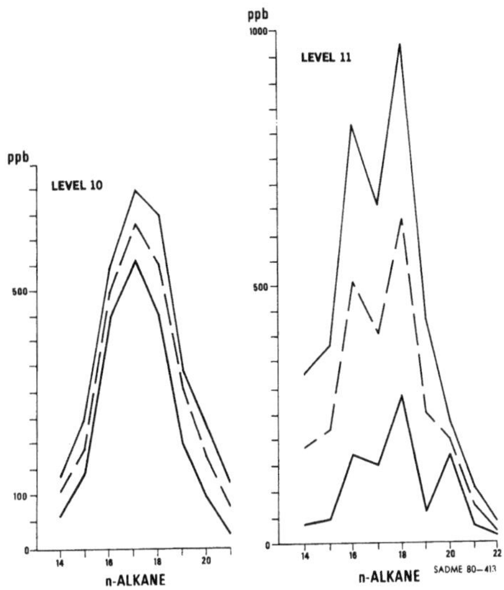

FIG. 16 - The distribution of n-alkanes in dolomites from the Malmani Dolomite, Western Deep Level Mine, Charletonville, Transvaal, South Africa (64). Dashed lines are average profiles.

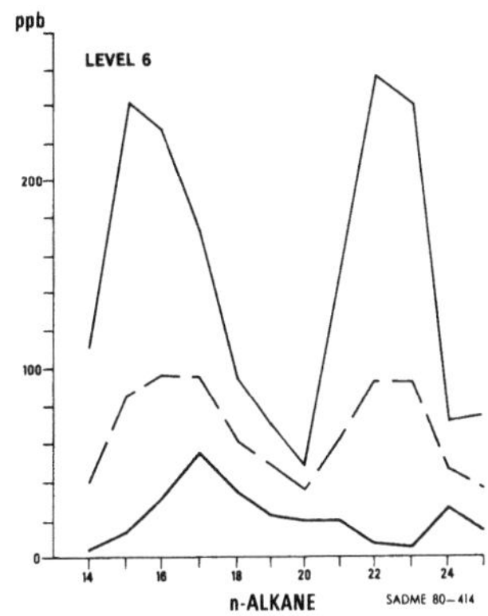

FIG. 17 - The distribution of n-alkanes in dolomites from Tweefontein, Transvaal, South Africa (64).

Comparison of the VOM in 2 Archaean stromatolitic limestones, one from the Bulawayan Group, Zimbabwe (Fig. 18), the other from the Insuzi Group, Pongola Supergroup, South Africa (not shown), reveals differences that are attributable to the lower metamorphic grade of the latter. Both n-alkane distributions are unimodal and show no odd or even carbon-number predominance.

However, in the Pongola sample, the n-alkane profile peaks at C_{20} and extends well beyond C_{23}; pr<ph, and pr/n-C_{17} and ph/n-C_{18} ratios are significantly higher than in the Bulawayan stromatolite.

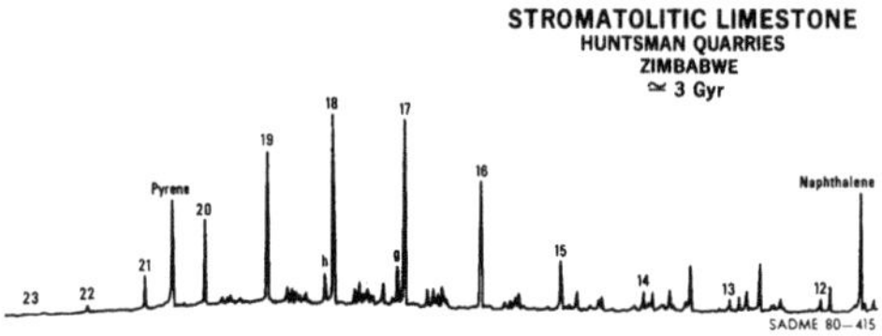

FIG. 18 - Gas chromatogram of volatile organic matter in the stromatolitic Bulawayan Limestone, Huntsman Quarry, Zimbabwe (64). Pyrene and naphthalene added as internal standards.

A study of the VOM in Brazilian BIF's (age 2.0-2.8 Gyr) showed that these ironstones are considerably richer in volatile organics than intercalated metaquartzites, which are essentially barren and about 5 times leaner than black shales within the sequence (16). The average ironstone n-alkane distribution differs little from the global average profile (Fig. 19), which in turn is reminiscent of the characteristic "algal" (including cyanobacterial) patterns (unimodal, maximum at or near n-C_{17}) seen in the Malmani Dolomite at Charletonville (Fig. 16). In complete contrast to the ironstone facies, the associated black shales gave a bimodal profile; the main maximum was at C_{11} or C_{12} and a second, less pronounced maximum at C_{17} or C_{18} (Fig. 19). The lower molecular-weight mode may be due to catagenetic breakdown products of algal lipids, represented by the primary (but now much diminished) mode between C_{16} and C_{19}. In the absence of clays this process would have been retarded in the ironstones; this may account for the higher value of the pr/n-C_{17} and ph/n-C_{18} ratios of these samples. Both shale and ironstone have pr/ph ratios in the range 0.7-3.0. The alkane distributions in the interbedded dolomites reflect those of the adjacent ironstone or black shale. The distribution of VOM in the Brazilian BIF's suggests that contamination of these rocks by epigenetic material is not significant.

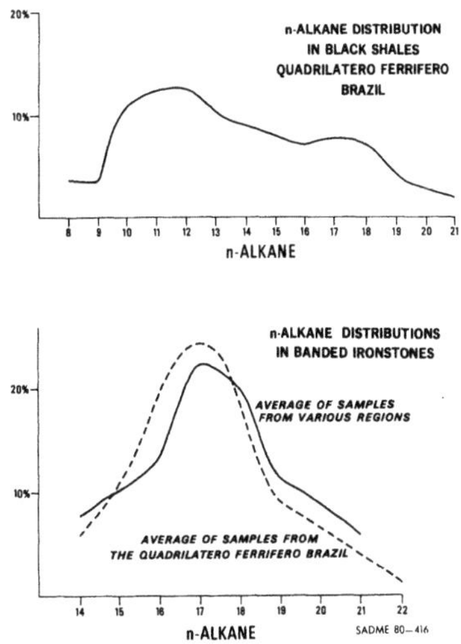

FIG. 19 - Distributions of n-alkanes in BIF's and black shales, Quadrilatero Ferrifero, Brazil. Average profile for BIF's based on samples from Brazil, Africa, India, N. America, and Russia(16).

A study of VOM in metasediments of the Isua Supracrustal Belt is still in progress. Volatile organics occur chiefly in quartzites and in silicate-rich BIF's. Alkane distributions show great variability, even over distances of a few meters. Some samples appear to be contaminated by organic matter of relatively recent origin (small n-alkane peaks on a huge unresolved naphthene hump, or humps), others by older epigenetic material of apparently higher-plant origin (wide distribution of n-alkanes with significant quantities and odd carbon-number predominance in the high molecular weight range). However, several rock samples have an alkane pattern similar to that of the Bulawayan stromatolitic limestone (see Fig. 18). These samples appear to contain syngenetic microbial organic matter at a rather late stage of catagenesis, an observation which agrees with the amphibolite-grade metamorphism of the Isua terrain.

CONCLUDING REMARKS

The occurrence of syngenetic kerogen and hydrocarbons dates back to the oldest known Precambrian sedimentary rocks. Their composition and concentration is governed by a complex interplay of variables that include the source biota, environment, rock type, and thermal history. Contamination by recent or epigenetic organic matter poses a serious problem for the "chemical fossil" approach to the study of early organic evolution. The problem is particularly acute in the case of hydrocarbons isolated from rocks which are lean in organics. Reliable chemical criteria for the syngeneity of Precambrian hydrocarbons are yet to be established. This will require a detailed and systematic understanding of the effects of catagenesis and metamorphism on organic matter in Precambrian sedimentary rock types. If the composition and concentration (per unit weight organic carbon) of the hydrocarbon fraction is broadly consistent with the depositional environment, lithofacies, kerogen type, and rank of its host rock, then it is probably both syngenetic and free of significant contamination.

The association of organic matter with Precambrian metallic ore deposits is relatively common (35,55) but inadequately documented, and its genetic implications are poorly understood. Where the absence of severe metamorphism provides a geochemical window through which the results of ancient ore-forming processes can be observed, the following palaeobiological and/or environmental indicators may be useful:

* kerogen type and the isotopic composition of carbon
* remnants of algal and bacterial lipids, pigments, carbohydrates, and proteins among the products of kerogen pyrolysis or ozonolysis
* carbon-number distribution and relative concentrations of normal, monomethyl, and acyclic isoprenoid alkanes in the geolipid extract or VOM

Organic matter in sedimentary ore deposits which formed in extreme environments may contain evidence of the unusual lipid system of archaebacteria, viz., a high relative abundance of C_{15}-C_{20} and C_{21+} isoprenoid alkanes (27,60), and an anomalous depletion of deuterium in the hydrocarbons and associated kerogen (25).

Acknowledgements. Supported in part by the Deutsche Forschungsgemeinschaft through its Sonderforschungsbereich 73 "Atmospheric Trace Components." D.M.M. publishes with the permission of the Director-General of Mines & Energy, South Australia.

REFERENCES

(1) Arefev, O.A.; Zabrodina, M.N.; Makushina, V.M.; and Petrov, A.A. 1980. Relic tetra- and pentacyclic hydrocarbons in the old oils of the Siberian Platform. Izv. Akad. Nauk. SSR, Ser. Geol. 3: 135-140 (in Russian).

(2) Awramik, S.M.; Schopf, J.W.; Walter, M.R.; and Buick, R. 1980. Filamentous fossil bacteria 3.5 Ga-old from the Archean of Western Australia. Science, in press.

(3) Barghoorn, E.S.; Knoll, A.H.; Dembicki, H., Jr.; and Meinschein, W.G. 1977. Variation in stable carbon isotopes in organic matter from the Gunflint Iron Formation. Geochim. Cosmochim. Acta 41: 425-430.

(4) Barghoorn, E.S.; Meinschein, W.G.; and Schopf, J.W. 1965. Paleobiology of a Precambrian shale. Science 148: 461-472.

(5) Bondesen, E.; Pedersen, K.R.; and Jørgensen, O. 1967. Precambrian organisms and the isotopic composition of organic remains in the Ketilidian of South-West Greenland. Medd. Grønland 179 (4).

(6) Brooks, J.D. 1971. Organic matter in Archaean rocks. Geol. Soc. Aust., Spec. Publs. 3: 413-418.

(7) Brooks, J., and Shaw, G. 1972. Geochemistry of sporopollenin. Chem. Geol. 10: 69-87.

(8) Craig, H. 1953. The geochemistry of the stable carbon isotopes. Geochim. Cosmochim. Acta 3: 53-92.

(9) Degens, E.T. 1969. Biogeochemistry of stable carbon isotopes. In Organic Geochemistry: Methods and Results, eds. G. Eglinton and M.T.J. Murphy, pp. 304-329. New York: Springer.

(10) Degens, E.T., and Mopper, K. 1976. Factors controlling the distribution and early diagenesis of organic material in marine sediments. In Chemical Oceanography, eds. J.P. Riley et al., vol. 6, pp. 59-113. New York: Academic Press.

(11) DeSouza, N.J., and Nes, W.R. 1968. Sterols: isolation from a blue-green alga. Science 162: 363.

(12) Dungworth, G., and Schwartz, A.W. 1974. Organic matter and trace elements in Precambrian rocks from South Africa. Chem. Geol. 14: 167-172.

(13) Eglinton, G., and Calvin, M. 1967. Chemical fossils. Sci. Amer. 216(1): 32-43.

(14) Eichmann, R., and Schidlowski, M. 1974. Isotopic composition of carbonaceous matter from the Precambrian uranium deposits of the Blind River district, Canada. Naturwissenschaften 61: 449.

(15) Eichmann, R., and Schidlowski, M. 1975. Isotopic fractionation between coexisting organic carbon - carbonate pairs in Precambrian sediments. Geochim. Cosmochim. Acta 39: 585-595.

(16) Fiebiger, W. 1975. Organische Substanzen in präkambrischen Itabiriten und deren Nebengesteinen. Geol. Rundschau 64: 641-652.

(17) Gavelin, S. 1957. Variations in isotopic composition of carbon from metamorphic rocks in northern Sweden and their geological significance. Geochim. Cosmochim. Acta 12: 297-314.

(18) Hahn, J. 1970. Eine gaschromatographische Methode zur schnellen Charakterisierung der organischen Substanz von Bohrkernen und anderen Sedimentproben. Erdöl und Kohle-Erdgas-Petrochemie 23: 790-792.

(19) Harrington, J.S., and Cilliers, J.J. le R. 1963. A possible origin of the primitive oils and amino acids isolated from amphibole asbestos and banded ironstone. Geochim. Cosmochim. Acta 27: 411-418.

(20) Hayes, J.M., and Wedeking, K.W. 1980. Organic geochemistry of Precambrian carbon. Interdisciplinary Study of the Origin and Evolution of Earth's Earliest Biosphere, Symposium Abstracts Vol., p. 7. University of California, Los Angeles.

(21) Hoefs, J., and Frey, M. 1976. The isotopic composition of carbonaceous matter in a metamorphic profile from the Swiss Alps. Geochim. Cosmochim. Acta 40: 945-951.

(22) Hoefs, J., and Schidlowski, M. 1967. Carbon isotope composition of carbonaceous matter from the Precambrian of the Witwatersrand System. Science 155: 1096-1097.

(23) Hoering, T.C. 1967. The organic geochemistry of Precambrian rocks. In Researches in Geochemistry, ed. P.H. Abelson, vol. 2., pp. 87-111. New York: Wiley.

(24) Hoering, T.C. 1976. Molecular fossils from the Precambrian Nonesuch Shale. Carnegie Inst., Wash., Yearbook 75: 806-813.

(25) Hoering, T.C. 1977. The stable isotopes of hydrogen in Precambrian organic matter. In Chemical Evolution of the Early Precambrian, ed. C. Ponnamperuma, pp. 81-86. New York: Academic.

(26) Hollerbach, A., and Welte, D.H. 1977. Über Sterane und Triterpane in Erdölen und ihre phylogenetische Bedeutung. Naturwissenschaften 64: 381-382.

(27) Holzer, G.; Oró, J.; and Tornabene, T.G. 1979. Gas chromatographic-mass spectrometric analysis of neutral lipids from methanogenic and thermoacidophilic bacteria. J. Chromatogr. 186: 873-887.

(28) Hunt, J.M. 1979. Petroleum Geochemistry and Geology. San Francisco: Freeman.

(29) Jackson, T.A.; Fritz, P.; and Drimmie, R. 1978. Stable carbon isotope ratios and chemical properties of kerogen and extractable organic matter in pre-Phanerozoic and Phanerozoic sediments - their interrelations and possible paleobiological significance. Chem. Geol. 21: 335-350.

(30) Johns, R.B.; Belsky, T.; McCarthy, E.D.; Burlingame, A.L.; Haug, P.; Schnoes, H.K.; Richter, W.; and Calvin, M. 1966. The organic geochemistry of ancient sediments - II. Geochim. Cosmochim. Acta 30: 1191-1222.

(31) Kaplan, I.R., and Nissenbaum, A. 1966. Anomalous carbon-isotope ratios in nonvolatile organic material. Science 153: 744-745.

(32) Knutson, J.; Ferguson, J.; Roberts, W.M.B.; Donelly, T.H.; and Lambert, I.B. 1979. Petrogenesis of the copper-bearing breccia pipes, Redbank, Northern Territory, Australia. Econ. Geol. 74: 814-826.

(33) Kvenvolden, K.A., and Roedder, E. 1971. Fluid inclusions in quartz crystals from South-West Africa. Geochim. Cosmochim. Acta 35: 1209-1229.

(34) Leventhal, J.; Suess, S.E.; and Cloud, P. 1975. Nonprevalence of biochemical fossils in kerogen from pre-Phanerozoic sediments. Proc. Nat. Acad. Sci. USA 72: 4706-4710.

(35) McKirdy, D.M. 1974. Organic geochemistry in Precambrian research. Precambrian Res. 1: 75-137.

(36) McKirdy, D.M. 1976. Biochemical markers in stromatolites. In Stromatolites, ed. M.R. Walter, pp. 163-191. Amsterdam: Elsevier.

(37) McKirdy, D.M. 1977. Diagenesis of microbial organic matter: a geochemical classification and its use in evaluating the hydrocarbon-generating potential of Proterozoic and Lower Palaeozoic sediments, Amadeus Basin, central Australia. Ph.D. Thesis, Geology Department, Australian National University, Canberra.

(38) McKirdy, D.M., and Kantsler, A.J. 1980. Oil geochemistry and potential source rocks of the Officer Basin, South Australia. Aust. Pet. Explor. Ass. J. 20(1): 68-86.

(39) McKirdy, D.M.; McHugh, D.J.; and Tardif, J.W. 1980. Comparative analysis of stromatolitic and other microbial kerogens by pyrolysis-hydrogenation-gas chromatography (PHGC). In Biogeochemistry of Ancient and Modern Environments, eds. P.A. Trudinger, M.R. Walter, and B.J. Ralph, pp. 187-200. Berlin: Springer-Verlag.

(40) McKirdy, D.M., and Powell, T.G. 1974. Metamorphic alteration of carbon isotopic composition in ancient sedimentary organic matter: new evidence from Australia and South Africa. Geology 2: 591-595.

(41) Meyerhoff, A.A. 1980. Geology and petroleum fields in Proterozoic and Lower Cambrian strata, Lena-Tunguska petroleum province Eastern Siberia, USSR. Amer. Ass. Pet. Geol. Memoir, in press.

(42) Minter, W.E.L. 1978. A sedimentological synthesis of placer gold, uranium and pyrite concentrations in Proterozoic Witwatersrand sediments. Can. Soc. Pet. Geol. Memoir 5: 801-829.

(43) Muir, M.D. 1978. Occurrence and potential uses of Archaean microfossils and organic matter. In Archaean Cherty Metasediments: Their Sedimentology, Micropalaeontology, Biogeochemistry, and Significance to Mineralization, eds. J.E. Glover and D.I. Groves, pp. 11-21. Publs. Geol. Dep. and Extension Service, University of Western Australia, 2.

(44) Nagy, B. 1976. Organic chemistry on the young earth: evolutionary trends between ~3,800 m.y. and ~2,300 m.y. ago. Naturwissenschaften 63: 499-505.

(45) Oehler, D.Z.; Schopf, J.W.; and Kvenvolden, K.A. 1972. Carbon isotopic studies of organic matter in Precambrian rocks. Science 175: 1246-1248.

(46) Oehler, D.Z., and Smith, J.W. 1977. Isotopic composition of reduced and oxidized carbon in Early Archaean rocks from Isua, Greenland. Precambrian Res. 5: 221-228.

(47) Pardue, J.W.; Scalan, R.S.; van Baalen, C.; and Parker, P.L. 1976. Maximum carbon isotope fractionation in photosynthesis by blue-green algae and a green alga. Geochim. Cosmochim. Acta 40: 309-312.

(48) Peat, C.J.; Muir, M.D.; Plumb, K.A.; McKirdy, D.M.; and Norvick, M.S. 1978. Proterozoic microfossils from the Roper Group, Northern Territory, Australia. Bur. Miner. Resour. J. Aust. Geol. Geophys. 3: 1-17.

(49) Perry, E.C., Jr., and Ahmad, S.N. 1977. Carbon isotope composition of graphite and carbonate minerals from 3.8 -AE metamorphosed sediments, Isukasia, Greenland. Earth Planet. Sci. Lett. 36: 280-284.

(50) Perry, E.C., Jr.; Tan, F.C.; and Morey, G.B. 1973. Geology and stable isotope geochemistry of the Biwabik Iron Formation, northern Minnesota. Econ. Geol. 68: 1110-1125.

(51) Powell, T.G.; Cook, P.J.; and McKirdy, D.M. 1975. Organic geochemistry of phosphorites: relevance to petroleum genesis. Amer. Ass. Pet. Geol. Bull. 59: 618-632.

(52) Powell, T.G.; Foscolos, A.E.; Gunther, P.R.; and Snowdon, L.R. 1978. Diagenesis of organic matter and fine clay minerals: a comparative study. Geochim. Cosmochim. Acta 42: 1181-1197.

(53) Reimer, T.O.; Barghoorn, E.S.; and Margulis, L. 1979. Primary productivity in an early Archaean microbial ecosystem. Precambrian Res. 9: 93-104.

(54) Reitz, R.C., and Hamilton, J.G. 1968. The isolation and identification of two sterols from two species of blue-green algae. Comp. Biochem. Physiol. 25: 401-416.

(55) Saxby, J.D. 1976. The significance of organic matter in ore genesis. In Handbook of Strata-bound and Stratiform Ore Deposits, ed. K.H. Wolf, pp. 111-133. Amsterdam: Elsevier.

(56) Schidlowski, M.; Appel, P.W.U.; Eichmann, R.; and Junge, C.E. 1979. Carbon isotope geochemistry of the 3.7 x 10^9-yr-old Isua sediments, West Greenland: implications for the Archaean carbon and oxygen cycles. Geochim. Cosmochim. Acta 43: 189-199.

(57) Sklarew, D.S., and Nagy, B. 1979. 2,5-Dimethylfuran from ~2.7 x 10^9-yr-old Rupemba-Belingwe stromatolite, Rhodesia: potential evidence for remnants of carbohydrates. Proc. Nat. Acad. Sci. USA 76: 10-14.

(58) Smith, J.W.; Schopf, J.W.; and Kaplan, I.R. 1970. Extractable organic matter in Precambrian cherts. Geochim. Cosmochim. Acta 34: 659-675.

(59) Tissot, B.P., and Welte, D.H. 1978. Petroleum Formation and Occurrence - a New Approach to Oil and Gas Exploration. Berlin: Springer-Verlag.

(60) Tornabene, T.G., and Langworthy, T.A. 1978. Diphytanyl and dibiphytanyl glycerol ether lipids of methanogenic archaebacteria. Science 203: 51-53.

(61) Walker, R.N.; Logan, R.G.; and Binnekamp, J.G. 1977. Recent geological advances concerning the H.Y.C. and associated deposits, McArthur River, N.T. J. Geol. Soc. Aust. 24: 365-380.

(62) Walter, M.R.; Buick, R.; and Dunlop, J.S.R. 1980. Stromatolites 3,400-3,500 Myr old from the North Pole area, Western Australia. Nature 284: 443-445.

(63) Welte, D.H., and Waples, D. 1973. Über die Bevorzugung geradzahliger n-Alkane in Sedimentgesteinen. Naturwissenschaften 60: 516-517.

(64) Wontka, J.F. 1979. Organisch-Geochemische Untersuchungen an präkambrischen Stromatolithen - Kalken und deren Nebengesteinen. Thesis, Department of Geosciences, Johannes-Gutenberg-University, Mainz, F.R. Germany.

(65) Ypma, P.J.M. 1979. Mineralogical and geological indications for the petroleum potential of the Etosha Basin, Namibia (S.W. Africa). Proc. Koninklijke Nederlandse Akademie van Wetenschappen, Series B, 82(1): 91-112.

(66) Zumberge, J.E.; Sigelo, A.C.; and Nagy, B. 1978. Molecular and elemental analyses of the carbonaceous matter in the gold and uranium bearing Vaal Reef carbon seams, Witwatersrand System. Min. Sci. Eng. 10: 223-246.

Mineral Deposits and the Evolution of the Biosphere, eds. H.D. Holland and M. Schidlowski, pp. 155-176. Dahlem Konferenzen, 1982.
Berlin, Heidelberg, New York: Springer-Verlag.

Prebiotic Synthesis of Organic Compounds

S. L. Miller
Dept. of Chemistry, University of California, San Diego
La Jolla, CA 92093, USA

Abstract. The heterotrophic hypothesis of the origin of life is now generally accepted. This involves the synthesis of simple organic compounds on the primitive Earth, the polymerization of these compounds, and the organization of the polymers into the first self-replicating organism. The need to synthesize simple organic compounds places constraints on conditions that prevailed on the primitive Earth. The temperature must have been low, O_2 was absent, and the atmosphere was reducing. The most effective atmosphere for the synthesis of organic compounds is CH_4, N_2 or NH_3, H_2O. Experiments with less reduced atmospheres such as CO, N_2 or NH_3, H_2O, H_2 and CO_2, N_2 or NH_3, H_2O, H_2 do give organic compounds but the yields are generally smaller and fewer compounds are obtained. CO and CO_2 atmospheres without H_2 give no organic compounds at all or very small yields. Prebiotic syntheses of amino acids, purines, pyrimidines, and sugars are now known. Organic compounds in carbonaceous chondrites are strikingly similar to those produced in laboratory syntheses with electric discharges.

INTRODUCTION

It is now generally accepted that life arose on the Earth early in its history. The sequence of events started with the synthesis of simple organic compounds by various processes. These simple organic compounds reacted to form polymers, which in turn reacted to form structures of progressively greater complexity, until a structure was formed which could be called living. This hypothesis is relatively new; it was first expressed clearly by Oparin (26) and has been elaborated by Haldane

(12), Urey (37), and Bernal (4). It is sometimes referred to as the Oparin-Haldane or the Heterotrophic hypothesis (14). Older ideas which are no longer regarded seriously include the seeding of the earth from another planet (panspermia), the origin of life at the present time from decaying organic material (spontaneous generation), and the origin of an organism early in the Earth's history by an extremely improbable event. In the last of these hypotheses it is assumed that conditions on Earth were essentially those of the present day, except possibly for the absence of molecular oxygen; the first organism would have to have been autotrophic, that is, it would have to have synthesized all its organic compounds from CO_2, H_2O, and light.

Oparin proposed that the first organisms were heterotrophic, i.e., that they used the organic compounds available in their environment. They still had to build compounds such as proteins and nucleic acids, but they did not have to synthesize amino acids, purines, pyrimidines, and sugars. Oparin, as well as Urey, also proposed that the Earth had an atmosphere consisting largely of CH_4, NH_3, H_2O, and H_2, and that organic compounds might have been synthesized in such an atmosphere. On the basis of Urey's and Oparin's ideas, it was shown that amino acids could be synthesized in surprisingly high yield by the action of electric sparks on strongly reducing atmospheres (22). There is now an extensive literature dealing with the prebiotic synthesis of organic compounds and polymers; since this literature is too extensive to be reviewed here, only its highlights will be covered (for reviews, see (15,19,24)). The heterotrophic hypothesis and the details of the pathways of prebiotic syntheses place constraints on geological conditions on the primitive Earth. Some of these constraints will be considered first.

The Time Available for the Origin of Life

The time period in which prebiotic synthesis of organic compounds took place is frequently misunderstood. The Earth is 4.5×10^9 years old, and the earliest fossil organisms known, the Warrawoona microfossils and stromatolites, are 3.5×10^9

years old. The difference is 1.0×10^9 years, but the time available for life to arise was probably shorter. It probably took a few hundred million years for organisms to evolve to the level of those found in the Warrawoona formation. In addition, if the Earth completely melted during its formation, then the time available would be further shortened by the time needed for the Earth to cool down sufficiently for organic compounds to be stable.

A period of say, 0.5×10^9 years does not, in my opinion, present any problems. Many writers have stressed that many improbable events were required for the origin of life, and therefore much time was needed. I believe that too much emphasis has been placed on the need for time. Periods of 10^9 years are so far removed from our experience that we have no feeling or judgement as to what is likely or unlikely in them. If the origin of life took only 10^6 years, I would not be surprised. It cannot be proved that 10^4 years is too short a period.

Melting of the Primitive Earth

There are many theories that seek to describe the events that took place during the condensation of cosmic dust and larger objects to form the Earth. At one extreme, it is believed that the accumulation of material took place in less than 10^5 years; in this scheme gravitational energy was also released sufficiently rapidly to melt the entire Earth, including material at its surface. At the other extreme, the accretion of material is believed to have been sufficiently slow, so that the gravitational energy released during accumulation was dissipated by radiation at a rate comparable to the rate of energy production. In this model the interior of the Earth would have melted due to the effects of adiabatic compression and the decay of radioactive elements. In both models a molten core would have formed.

Whether the entire Earth was ever molten or not does not greatly affect our discussion of the prebiotic synthesis of organic compounds. If the entire Earth did melt, then all organic compounds, both in the interior and on the surface, would have been pyrolyzed completely to an equilibrium mixture consisting largely of CO_2, CO, CH_4, H_2, N_2, NH_3, and H_2O. All organic compounds synthesized in the solar nebula and reaching the surface of the Earth intact would have been destroyed on a molten Earth. When the Earth had cooled down sufficiently, a crust would have formed. When the average temperature on the surface and in the atmosphere became low enough, organic compounds would have been synthesized and, most important, would have accumulated. If the Earth was never completely molten, the prebiotic synthesis of organic compounds began after most of the Earth had accumulated. It is possible that some of the organic compounds synthesized in the solar nebula were brought to the Earth with dust particles, meteorites, and comets and that they survived their impact with the atmosphere and with the Earth's surface. It is not clear how much of a contribution these compounds made to the inventory of prebiotic organic compounds. Relatively unstable compounds, such as sugars and certain amino acids, cannot be accounted for in this manner since they would have decomposed before life could arise. There would have been a continuous addition of carbonaceous chondrites after the initial major accretion, but they would not result in the accumulation of unstable organic compounds. In any case, unstable organic compounds such as sugars have not been found in carbonaceous chondrites. Thus for unstable compounds a continuous synthesis is required, which means synthesis in the atmosphere and oceans of the primitive Earth.

Temperature of the Primitive Earth

The necessity to accumulate organic compounds for the synthesis of the first organism requires that the temperature of the Earth was reasonably low; otherwise the required organic compounds would have decomposed. The half-life for decomposition varies from several billion years for alanine at 25°, to a few million years for serine, 10^3 to 10^5 years for the hydrolysis of peptides and polynucleotides, and at most a few hundred years for sugars. Since the temperature coefficient of the rate of decomposition of organic compounds tends to be large, the half-life of these compounds would be much less at 50° or 100°C. Conversely, the half-lives would be considerable longer at 0°C. The rates of hydrolysis of peptide and polynucleotide polymers and the rate of decomposition of sugars are so large that it seems impossible for such compounds to have accumulated sufficiently in aqueous solutions for use in the first organism, unless surface temperatures were low.

The average temperature of the present ocean is 4°C; surface waters are somewhat warmer. While the freezing point of pure water is 0°C, seawater begins to freeze at -1.8°C. Seawater is almost completely frozen at -21°C. The temperature of the primitive oceans is not known, but the instability of organic compounds and polymers is such that life could probably not have arisen in the ocean unless its mean temperature was below 25°C. A temperature of 0°C would have helped greatly, and -21°C would have been even better. At such low temperatures, most of the water on the primitive Earth would have been in the form of ice; liquid seawater would have been confined to the equatorial oceans.

There is another reason for believing that life evolved at low temperatures. All of the template-directed reactions that must have led to the emergence of biological organization take place only below the melting temperature of the appropriate organized polynucleotide structure. These temperatures range from 0°C, or lower, to perhaps 35°C, in the case of polynucleotide-mononucleotide helices.

The environment in which life arose is frequently referred to as a warm, dilute soup of organic compounds. I believe that a cold, concentrated soup would have provided a better environment for the origins of life. At first sight low temperatures might seem to have been a disadvantage, because chemical syntheses would have proceeded more slowly. However, if ample time was available, it was the ratio of the rates of synthesis to the rates of decomposition which were important, rather than the absolute rates themselves. Since the temperature coefficients of the synthesis reactions are generally less than those for the decomposition reactions, low temperatures would have favored the synthesis of complex organic compounds and polymers.

Oxygen in the Primitive Atmosphere

Molecular oxygen is usually assumed to have been absent from the primitive atmosphere. There are several reasons for this belief. In the absence of O_2-evolving photosynthetic organisms, the photodissociation of water in the upper atmosphere would have been the major source of O_2. The small amount of O_2 produced by this mechanism would probably have been removed by reaction with Fe^{+2} and other reduced inorganic compounds. If free O_2 had been present, it would have reacted relatively rapidly with organic compounds, especially in the presence of ultraviolet light. These and related arguments are so compelling that it does not seem possible that organic compounds were present in the primitive ocean for any length of time after large amounts of O_2 had accumulated in the Earth's atmosphere. Today, organic compounds are present at the surface of the Earth only because they are being synthesized continuously by living organisms. Organic compounds occur below the surface of the Earth, for example, in coal and oil, because there the environment is anaerobic. It appears certain, then, that O_2 in substantial amounts was absent from the Earth's atmosphere during the period when organic compounds were synthesized and probably up to the time when the first organism evolved.

A corollary of these arguments is that if life does not develop on a planet by the time its atmosphere becomes oxidizing, then the organic compounds would be decomposed thermally or by oxidation with O_2, and life will never arise on that planet unless reducing conditions are reestablished at some later time.

The Composition of the Primitive Atmosphere

There is no agreement on the constituents of the primitive atmosphere. There is presently no geological evidence concerning the conditions on the Earth between 4.5 x 10^9 years and 3.8 x 10^9 years since no rocks older than 3.8 x 10^9 years are known. Even the 3.8 x 10^9 year old metasediments at Isua are too highly metamorphosed to have preserved detailed information regarding the composition of the atmosphere at that time. The proposed compositions of the early atmosphere range from strongly reducing (CH_4, NH_3, H_2O, H_2) to nonreducing (CO_2, N_2, H_2O). The various proposed compositions and the reasons why they have been put forward will not be discussed here. As will be shown in the next section, the yields and the variety of organic compounds obtained abiotically are considerably greater in the more reducing atmospheres, and some of the required prebiotic organic chemistry sets rather explicit restrictions on the abundance of certain atmospheric constituents. Such considerations cannot prove that the earth had a certain primitive atmosphere, but the prebiotic synthesis constraints should be a major consideration.

PREBIOTIC SYNTHESES

Energy Sources

A wide variety of energy sources and gas mixtures have been used since the first experiments in prebiotic chemistry with electric discharges. The importance of a given energy source is determined by the product of the energy available and its efficiency for the synthesis of organic compounds. These factors cannot be evaluated with precision, but a qualitative assessment of the energy sources can be made. It should be emphasized that a single source of energy or a single process is unlikely to have produced all the organic compounds on the primitive Earth. An estimate of the sources of energy on the Earth at the present time is given in Table 1 (taken from (25)).

The energy from the decay of radioactive elements was probably not an important energy source for the synthesis of organic compounds on the primitive Earth since most of the ionization

TABLE 1 - Present sources of energy averaged over the earth.

Source	Energy (cal cm^{-2} yr^{-1})
Total radiation from sun	260,000
Ultraviolet light	
<3000 Å	3,400
<2500 Å	563
<2000 Å	41
<1500 Å	1.7
Electric discharges	4
Cosmic rays	0.0015
Radioactivity (to 1.0 km depth)	0.8
Volcanoes	0.13
Shock waves	1.1
Solar wind	0.2

due to radioactive decay took place in silicate rocks rather than in the atmosphere. The shock wave energy generated by the impact of meteorites on the Earth's atmosphere and surface and the larger quantity of shock wave energy generated in lightning bolts have been proposed as important energy sources for primitive Earth organic synthesis. Very high yields of amino acids have been reported in some experiments (2), but it is doubtful whether such yields would be obtained in natural shock waves. Cosmic rays are a minor source of energy on the Earth at present, and it seems unlikely that they have ever been a major source of energy.

The thermal energy in the lava emitted at the present time is a significant but not a major source of energy. It is generally supposed that there was a much greater amount of volcanic activity on the primitive Earth, but there is no direct evidence to support this. Even if volcanic activity was ten times more intense than at present, it was probably not the dominant energy source. Nevertheless, molten lava may have been important in the pyrolytic synthesis of organic compounds (see below).

Ultraviolet light was probably the largest source of energy on the primitive Earth. With the exception of ammonia (<2300 Å) and H_2S ($<$2600 Å), all of the likely constituents of the early atmosphere absorbed ultraviolet light of wavelength below 2000 Å. Whether UV light was the most effective energy source for the synthesis of organic compounds is not clear. Most of the photochemical reactions would have occurred in the upper atmosphere; the products formed there would, for the most part, have absorbed at longer wavelengths and would have been decomposed before reaching the protection of the oceans. The yield of amino acids from the photolysis of CH_4, NH_3, and H_2O at wavelengths of 1470 and 1294 Å is quite low (11), probably due to the low yield of hydrogen cyanide. The yield of amino acids in the photolysis of mixtures of CH_4, C_2H_6, NH_3, H_2O, and H_2S by UV light of wavelength greater than 2000 Å (32) is also low, but the amount of energy in this part of the solar spectrum is much greater. H_2S is the only component of such mixtures to absorb ultraviolet light of wavelength $>$ 2000 Å, but the photodissociation of H_2S produces hydrogen atoms having a high kinetic energy and these H atoms can activate or dissociate methane, ammonia, and water. This appears to be a very attractive pathway for prebiotic syntheses. However, it is not certain that enough H_2S could have been maintained in the early atmosphere, since H_2S is photolyzed rapidly to elemental sulfur and hydrogen.

The most widely used sources of energy for laboratory syntheses of prebiotic compounds are electric discharges. These include sparks, semi-corona, arc, and silent discharges; among these, spark sources have been used most frequently. Their ease of handling and high efficiency have been factors favoring their use, but the most important reason for their use is that electric discharges are a very efficient means for synthesizing hydrogen cyanide, while ultraviolet light is not. Hydrogen cyanide is a central intermediate in prebiotic syntheses; it is needed for amino acid synthesis via the Strecker reaction, or via self-polymerization to amino acids, and - most importantly - for the prebiotic synthesis of adenine and guanine.

An important feature of all laboratory syntheses using these energy sources is the activation of molecules in a local area followed by quenching of the activated mixture and the protection of the organic products from the disruptive influence of the energy sources. Quenching and protection are critical; in their absence the organic products are destroyed by continuous exposure to the energy source.

Prebiotic Synthesis of Amino Acids

Mixtures of CH_4, NH_3, and H_2O with or without H_2 are considered strongly reducing atmospheres. The atmosphere of Jupiter contains these species, with H_2 in large excess over the CH_4. The first successful prebiotic amino acid synthesis was carried out using this gas mixture and an electric discharge as an energy source (22). The result was a large yield of amino acids (the yield of glycine alone was 2.1% based on the carbon), together with hydroxy acids, short chain aliphatic acids, and urea. Surprisingly, the products were not a random mixture of organic compounds; rather a relatively small number of compounds were produced in substantial yield, and these compounds were, with few exceptions, of biological importance.

The mechanism of the synthesis of the amino and hydroxy acids in these experiments has since been investigated (23). It has been shown that the amino acids were not formed directly in the electric discharge but were the result of reactions in aqueous solution of smaller molecules produced in the discharge - in particular hydrogen cyanide and aldehydes. The reactions are

$$RCHO + HCN + NH_3 \rightleftarrows RCH(NH_2)CN \xrightarrow{H_2O} RCH(NH_2)\overset{O}{\overset{\|}{C}}-NH_2 \xrightarrow{H_2O} RCH(NH_2)COOH$$

$$RCHO + HCN \rightleftarrows RCH(OH)CN \xrightarrow{H_2O} RCH(OH)\overset{O}{\overset{\|}{C}}-NH_2 \xrightarrow{H_2O} RCH(OH)COOH$$

Subsequently, these reactions were studied in detail, and their equilibrium and rate constants were measured (39). The results showed that amino and hydroxy acids could have been synthesized at high dilutions ($\sim 10^{-7}$ M for alanine at 0°, with higher concentrations needed at higher temperature) of HCN and aldehydes in a primitive ocean. The rates of these syntheses are rather rapid. The half-lives for the hydrolysis of alanine nitrile and lactonitrile, the rate limiting steps in the synthesis of alanine and lactic acid, are 50 years and 2000 years at $0^{o}C$.

This synthesis of amino acids, called the Strecker Synthesis, requires the presence of NH_4^+ (and NH_3) in the primitive ocean. On the basis of the experimentally determined equilibrium and rate constants for glycine and alanine nitriles, it can be shown that equal amounts of amino and hydroxy acids are obtained when the NH_4^+ concentration is about 0.01 M at pH 8 and 25°; the required NH_4^+ concentration is rather insensitive to changes in temperature and pH. The NH_3 pressure in the atmosphere in equilibrium with an NH_4^+ concentration of 0.01 M at a pH of 8 is 2×10^{-7} atm at 0° and 4×10^{-6} atm at 25°. These are low partial pressures of NH_3, but they are necessary for amino acid synthesis. Ammonia is decomposed by ultraviolet light, but mechanisms for its resynthesis are available. However, the details of the ammonia balance on the primitive Earth remain to be worked out.

In a typical electric discharge experiment, the partial pressure of CH_4 is 0.1 to 0.2 atm. This pressure is used for the sake of convenience, and it is likely, but has never been demonstrated, that organic compounds could be synthesized at much lower partial pressures of methane. The constraints on pCH_4 on the primitive Earth are weak, but 10^{-5} to 10^{-3} atm seems a reasonable range. Higher CH_4 pressures are unlikely, because CH_4 would probably have been converted to a variety of organic compounds and removed into the oceans too rapidly for higher pressures of CH_4 to build up.

Ultraviolet light acting on a mixture of CH_4, NH_3, and H_2O is not effective in producing amino acids except at very short wavelengths (<1500 Å) and even then the yields are very low (11). The low yields are probably due to the low yields of HCN produced by ultraviolet light. If the gas mixture is modified by adding gases such as H_2S or formaldehyde, then reasonable yields of amino acids can be obtained at relatively long wavelengths (<2500 Å) where considerable quantities of solar energy are available (32). H_2S absorbs UV light at the longer wavelengths and is photodissociated to H and HS. The H atoms have a sufficiently high velocity to dissociate CH_4 and NH_3. It is possible, but has not been demonstrated, that HCN and other molecules are produced, which then form amino acids in the aqueous part of the system.

The pyrolysis of CH_4 and NH_3 between 800° and $1200^{\circ}C$ at contact times of a second or less gives very low yields of amino acids. However, the pyrolysis of CH_4 and other hydrocarbons gives good yields of benzene, phenylacetylene, and many other hydrocarbons. It can be shown that phenylacetylene would have been converted to phenylalanine and tyrosine in the primitive ocean (7). Pyrolysis of the hydrocarbons in the presence of NH_3 gives substantial yields of indole, which could have been converted to tryptophan in the primitive ocean.

A mixture of CH_4, N_2 with traces of NH_3, and H_2O is a more realistic atmosphere for the primitive earth because large amounts of NH_3 would not have accumulated in the atmosphere because the NH_4 would dissolve in the ocean. It is still, however, a strongly reducing atmosphere. This mixture of gases is quite effective with an electric discharge in producing amino acids (31,40). The yields are somewhat lower than with mixtures which are richer in NH_3, but the products are more diverse (Table 2). Hydroxy acids, short aliphatic acids, and dicarboxylic acids have been produced along with the amino acids. Ten of the 20 amino acids that occur in proteins were produced directly in the experiment summarized in Table 2. Counting asparagine and glutamine, which are formed but hydrolyzed before analysis, and methionine, which is formed when H_2S is added, one can say that 13 of the 20 amino acids in proteins formed in this experiment. Cysteine was found in the photolysis products of CH_4, NH_3, H_2O, and H_2S (32). The pyrolysis of hydrocarbons, as discussed above, leads

TABLE 2 - Yields from sparking CH_4 (PCH_4 = 200 nn; 336 m moles), N_2, and H_2O with traces of NH_3.

	μmole		μmole
Glycine	440	α, γ-Diaminobutyric acid	33
Alanine	790	α-Hydrocy-γ-aminobutyric acid	74
α-Amino-n-butyric acid	270	α, β-Diaminopropionic	6.4
α-Aminoisobutyric acid	~ 30	Isoserine	5.5
Valine	19.5	Sarcosine	55
Norvaline	61	N-Ethylglycine	30
Isovaline	~ 5	N-Propylglycine	~ 2
Leucine	11.3	N-Isopropylglycine	~ 2
Isoleucine	4.8	N-Methylalanine	~ 15
Alloisoleucine	5.1	N-Ethylalanine	< 0.2
Norleucine	6.0	β-Alanine	18.8
tert-Leucine	< 0.02	β-Amino-n-butyric acid	~ 0.3
Proline	1.5	β-Amino-isobutyric acid	~ 0.3
Aspartic acid	34	γ-Aminobutyric acid	2.4
Glutamic acid	7.7	N-Methyl-β-alanine	~ 5
Serine	5.0	N-Ethyl-β-alanine	~ 2
Threonine	~ 0.8	Pipecolic acid	~ 0.05
Allothreonine	~ 0.8		

Yield based on the carbon added as CH_4. Glycine = 0.26%, Alanine = 0.71%, total yield of amino acids in the table = 1.90%.

to phenylalanine, tyrosine, and tryptophan. This leaves the basic amino acids - lysine, arginine, and histidine - for which there are no established prebiotic syntheses. This problem may be solved before too long.

It can be shown that HCN and various aldehydes are produced by sparking mixtures of CH_4, N_2, NH_3, and H_2O. The most effective way to study these systems is to analyze the products from sparking mixtures of CH_4 + NH_3, CH_4 + N_2, and CH_4 + H_2O separately. The results show that nitriles, aldehydes, ketones, alcohols, and hydrocarbons are produced. Hydrogen cyanide is probably the most important product, since it was probably involved in the synthesis of both amino acids and purines in the primitive ocean. Other products include cyanoacetylene, which gives pyrimidines and aspartic acid, acrolein, which gives various amino acids including methionine, propiolaldehyde which gives nicotinamide, and cyanamide and cyanogen which are dehydrating (polymerizing) agents.

Mildly Reducing and Nonreducing Atmospheres

There has been less experimental work with gas mixtures containing CO and CO_2 rather than CH_4, and the experiments carried out so far have produced lower yields and a smaller variety of organic compounds ((1), and Miller et al., unpublished results).

Electric discharges acting on a mixture of CO, N_2, H_2 are not effective in amino acid synthesis unless the ratio of H_2 to CO is greater than 1. Glycine is produced in fair yield, but only small amounts of higher amino acids are produced. Large amounts of formaldehyde are obtained, however, and formaldehyde is important in the prebiotic synthesis of sugars.

Mixtures of CO_2, N_2, and H_2 are more oxidized than corresponding mixtures with CO. The yields of amino acids are quite low in electric discharge experiments unless the H_2/CO_2 ratio is greater than about 2. In that case glycine is produced in fair yield, but very little of the higher amino acids is formed.

Mixtures of CO + H_2 are used in the Fischer-Tropsch reaction to make hydrocarbons in high yields. The reaction requires a catalyst, usually Fe or Ni supported on silica, a temperature of 200-400°C, and a short contact time. Under various conditions, aliphatic hydrocarbons, aromatic hydrocarbons, alcohols, and acids can be produced. If NH_3 is added to CO + H_2 mixtures, amino acids, purines, and pyrimidines can be formed (13,41). The intermediates in these reactions are not known, but it is likely that HCN is produced together with other intermediates involved in electric discharge processes.

Mixtures of CO + H_2O are not effective in organic compound syntheses with electric discharges, but ultraviolet light that is absorbed by water (<1849 Å) results in the production of formaldehyde and other aldehydes, alcohols, and acids in fair yields (3). The mechanism seems to involve splitting H_2O to H + OH; OH radicals then convert CO to CO_2, and the H radicals reduce the CO.

Neither electric discharges nor ultraviolet light have yielded organic compounds when they act on mixtures of CO_2 + H_2O.

Ionizing radiation (e.g., 40 MeV helium ions) does give small yields of formic acid and formaldehyde (10).

The action of γ-rays on aqueous solutions of CO_2 and ferrous ion gives fair yields of formic acid, oxalic acid, and other simple products. Irradiation of such solutions with ultraviolet light gives similar results (9). In these reactions, the Fe^{+2} is a stoichiometric reducing agent rather than a catalyst. Nitrogen in the form of N_2 does not react, and experiments with NH_3 have not been tried.

Purine and Pyrimidine Syntheses

Hydrogen cyanide is used in the synthesis of purines as well as amino acids. This is illustrated in a remarkable synthesis of adenine. If strongly ammoniacal solutions of hydrogen cyanide are refluxed for a few days, adenine is obtained in up to 0.5 percent yield together along with 4-aminoimidazole-5-carboxamide and the usual cyanide polymer (27-29).

The mechanism of adenine synthesis in these experiments is probably

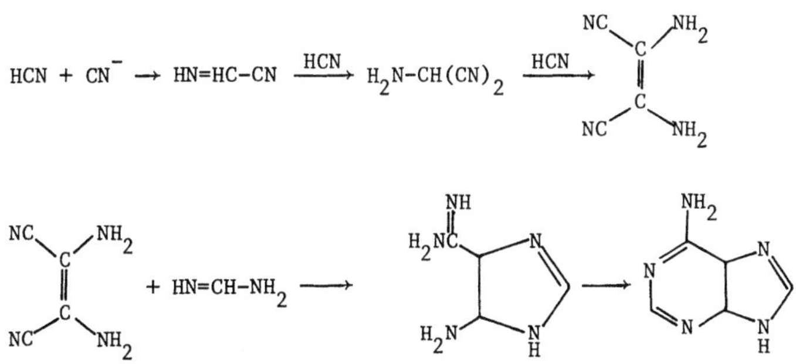

The difficult step in this synthesis of adenine is the reaction of the tetramer with formamidine. This step may be bypassed by the photochemical rearrangement of tetramer to aminoimidazole carboxamide, a reaction that proceeds readily in contemporary sunlight (34,35).

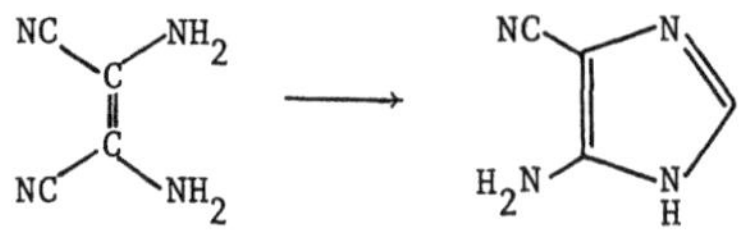

Tetramer formation may also have occurred in a eutectic solution. High yields of tetramer (>10 percent) can be obtained by cooling dilute cyanide solutions to between -10°C to -30°C for a few months.

The prebiotic synthesis of the pyrimidine cytosine involves cyanoacetylene, which is synthesized in good yield by sparking mixtures of CH_4 + N_2. Cyanoacetylene reacts with cyanate to give cytosine (33), which can be converted to uracil. Cyanate can come from cyanogen or from the decomposition of urea.

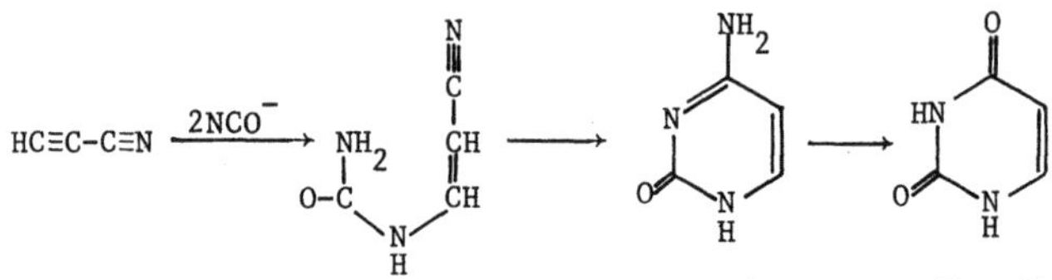

Cytosine Uracil

An alternative prebiotic synthesis of uracil starts with β-alanine and involves cyanate and ultraviolet light (36).

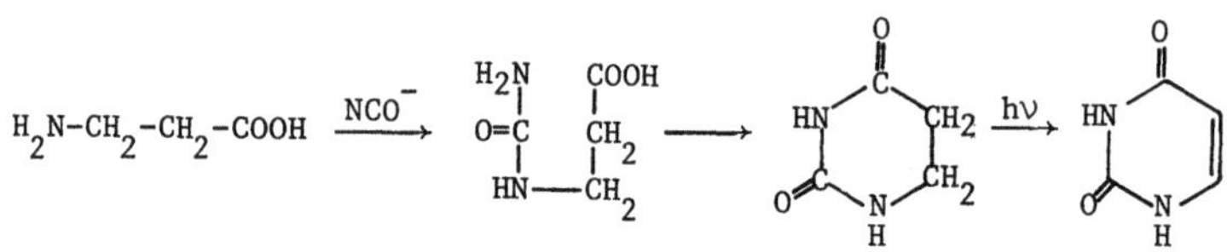

Sugars

The synthesis of reducing sugars from formaldehyde under alkaline conditions was discovered long ago. However, the process is very complex and is still understood incompletely. In simple solutions, the synthesis depends on the presence of a suitable catalyst. Calcium hydroxide and calcium carbonate are among the most popular of these. In the absence of catalysts, little or no sugar is obtained. At 100°C clays such as kaolin serve to catalyze the formation of monosaccharides, including ribose, in good yield from dilute (0.01 M) solutions of formaldehyde (8,30). The reaction is autocatalytic and proceeds in stages through glycolaldehyde, glyceraldehyde, dihydroxyacetone, tetroses and pentoses to hexoses that include glucose and fructose. One proposed reaction sequence is

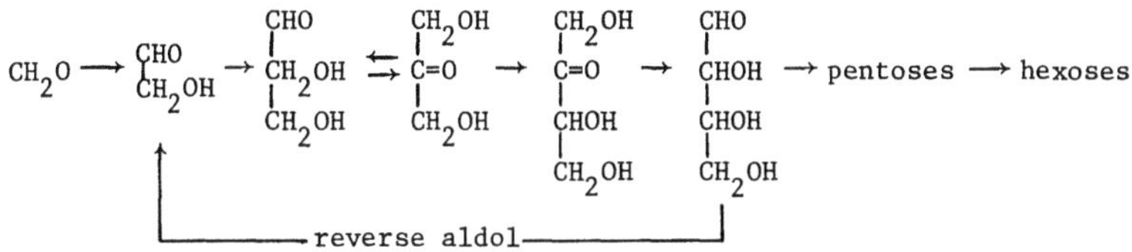

The problem with sugars on the primitive Earth is not their synthesis, but their stability. They decompose in a few hundred years or less at $25^{\circ}C$. There are a number of ways to stabllize sugars; the most interesting of these is the conversion of sugars to glycosides of a purine or pyrimidine.

<u>Polymer Synthesis and Organization</u>

A thorough discussion of this topic is beyond the scope of the present review. Suffice it to mention that there are two types of polymerization processes. The first uses a "high energy" compound such as cyanamide, cyanate, or polyphosphate. Such syntheses work, but the yields tend to be low. The second process involves heating dry mixtures to temperatures of 60° to $200^{\circ}C$. Only temperatures from 60° to 100° are plausible for prebiotic syntheses since the higher temperatures would decompose the organic compounds under geological conditions. These syntheses have been reviewed (21).

A number of attempts have been made to carry out nonenzymatic, template-directed syntheses of nucleic acids. There are a number of problems with such reactions, but the recent success of experiments in which ZN^{+2} was used as a catalyst (20) suggests that considerable advances in this direction can be made.

Organic Compounds in Carbonaceous Chondrites

On September 28, 1969, a type II carbonaceous chondrite fell in Murchison, Australia. Surprisingly large amounts of amino acids were found in this meteorite by Kvenvolden et al. (16,17). The first report identified seven amino acids (glycine, alanine, valine, proline, glutamic acid, sarcosine, and α-aminoisobutyric acid), of which all but valine and proline had been found in the original electric discharge experiments (22,23). The most striking of these are sarcosine and α-aminoisobutyric acid. The second report identified 18 amino acids of which nine had prevously been identified in the products of the electric discharge experiments.

At that time we had identified the hydrophobic amino acids from the low temperature electric discharge experiments described above; we therefore examined the products for the nonprotein amino acids found in Murchison. We were able to find all of them (31,40). There is a striking similarity between the products and relative abundances of the amino acids produced in experiments with electric discharges and those found in the Murchison meteorite. Table 3 compares the results. The most notable difference between amino acids in the Murchison meteorite

TABLE 3 - Relative abundances of amino acids in the Murchison meteorite and in an electric discharge synthesis (Table 2). Mole ratio to glycine (100):0.05-0-0.5,*; 0.5-5,**; 5-50***, >50****.

Amino acid	Murchison meteorite	Electric discharge
Glycine	****	****
Alanine	****	****
α-Amino-n-butyric acid	***	****
α-Aminoisobutyric acid	****	**
Valine	***	**
Norvaline	***	***
Isovaline	**	**
Proline	***	*
Pipecolic acid	*	<*
Aspartic acid	***	***
Glutamic acid	***	**
β-Alanine	**	**
β-Amino-n-butyric acid	*	*
β-Aminoisobutyric acid	*	*
γ-Aminobutyric acid	*	**
Sarcosine	**	***
N-Ethylglycine	**	***
N-Methylalanine	**	**

and those produced in the electric discharge experiments is the very low yield of pipecolic acid in the electric discharge experiments. Proline is also produced in relatively low yield by an electric discharge. The amount of α-aminoisobutyric acid is greater than α-amino-n-butyric acid in the meteorite, but the reverse was found in the electric discharge experiments. We do not believe that these differences in the ratio of the amino acids detract significantly from the overall similarity, especially since the ratio of α-aminoisobutyric acid to glycine is quite different in two meteorites of the same type; it is 0.4 in Murchison and 3.8 in Murray (5). A similar comparison has also been made between the dicarboxylic acids in Murchison (18) and those produced by an electric discharge (42); the product ratios are quite similar.

The close correspondence between the amino acids found in the Murchison meteorite and those produced by an electric discharge synthesis suggests that the amino acids in the meteorite were synthesized on the parent body of this meteorite by means of an electric discharge or by analogous processes. Electric discharges appear to be the most likely source of energy, but sufficient data are not available to make a realistic comparison with the efficacy of other energy sources. In any case, it is unlikely that a single source of energy was responsible for the synthesis of all of the organic compounds either on the parent body of the carbonaceous chondrites or on the primitive Earth. All sources of energy would have made their contribution, and the problem is to evaluate the relative importance of each source.

Our ideas on the prebiotic synthesis of organic compounds are based largely on the results of experiments in model systems. It is extremely gratifying to find that syntheses of this type apparently occurred on the parent body of at least some meteorites; this finding increases the likelihood that similar syntheses occurred on the primitive Earth.

Acknowledgement. This work has been supported bv NASA Grant No. NAGW-20.

REFERENCES

(1) Abelson, P.H. 1965. Abiogenic synthesis in the Martian environment. Proc. Natl. Acad. Sci. USA 54: 1490-1494.

(2) Bar-Nun, A.; Bar-Nun, N.; Bauer, S.H.; and Sagan, C. 1970. Shock synthesis of amino acids in simulated primitive environments. Science 168: 470-473.

(3) Bar-Nun, A., and Hartman, H. 1978. Synthesis of organic compounds from carbon monoxide and water by UV photolysis. Origins of Life 9: 93-101.

(4) Bernal, J.D. 1951. The Physical Basis of Life. London: Routledge and Kegan Paul.

(5) Cronin, J.R., and Moore, C.B. 1971. Amino acid analyses of the Murchison, Murray, and Allende carbonaceous chondrites. Science 172: 1327-1329.

(6) Ferris, J.P.; Sanchez, R.A.; and Orgel, L.E. 1968. Synthesis of pyrimidines from cyanoacetylene and cyanate. J. Mol. Biol. 33: 693-704.

(7) Friedmann, N., and Miller, S.L. 1969. Phenylalanine and tyrosine synthesis under primitive earth conditions. Science 166: 766-767.

(8) Gabel, N.W., and Ponnamperuma, C. 1967. Model for the origin of monosaccharides. Nature 216: 453-455.

(9) Getoff, N. 1962. Reduktion der Kohlensäure in wässeriger Lösung unter Einwirkung von UV-licht. Z. Naturforsch. 17b: 87-90, 751-757.

(10) Garrison, W.M.; Morrison, D.C.; Hamilton, J.G.; Benson, A.A.; and Calvin, M. 1951. Reduction of carbon dioxide in aqueous solutions by ionizing radiation. Science 114: 416-418.

(11) Groth, W., and v. Weyssenhoff, H. 1960. Photochemical formation of organic compounds from mixtures of simple gases. Planet. Space Sci. 2: 79-85.

(12) Haldane, J.B.S. 1929. Rationalist Annual 148: 3; reprinted In Science and Human Life. New York and London: Harper Bros., 1933, p. 149.

(13) Hayatsu, R., et al. 1972. Catalytic synthesis of nitriles, nitrogen bases and porphyrin-like pigments. Geochim. Cosmochim. Acta 36: 555-571.

(14) Horowitz, N.H. 1945. On the evolution of biochemical synthesis. Proc. Natl. Acad. Sci. USA 31: 153-157.

(15) Kenyon, D.H., and Steinmann, G. 1969. Biochemical Predestination. New York: McGraw-Hill.

(16) Kvenvolden, K.; Lawless, J.G.; Pering, K.; Peterson, E.; Flores, J.; Ponnamperuma, C.; Kaplan, I.R.; and Moore, C. 1970. Evidence for extraterrestrial amino-acids and hydrocarbons in the Murchison Meteorite. Nature 228: 923-926.

(17) Kvenvolden, K.A.; Lawless, J.G.; and Ponnamperuma, C. 1971. Nonprotein amino acids in the Murchison Meteorite. Proc. Natl. Acad. Sci. USA 68: 486-490.

(18) Lawless, J.G.; Zeitman, B.; Pereira, W.E.; Summons, R.E.; Duffield, A.M. 1974. Dicarboxylic acids in the Murchison Meteorite. Nature 251: 40-42.

(19) Lemmon, R.M. 1970. Chemical evolution. Chem. Rev. 70: 95-109.

(20) Lohrmann, R.; Bridson, P.K.; and Orgel, L.E. 1980. Efficient metal-ion catalyzed template-directed oligonucleotide synthesis. Science 208: 1464-1465.

(21) Lohrmann, R., and Orgel, L.E. 1973. Prebiotic activation processes. Nature 224: 418-420.

(22) Miller, S.L. 1953. A production of amino acids under possible primitive earth conditions. Science 117: 528-529.

(23) Miller, S.L. 1957. The formation of organic compounds on the primitive earth. Ann. NY Acad. Sci. 69: 260-274; reprinted In The Origin of Life on the Earth, ed. A. Oparin. Oxford: Pergamon Press, 1959, pp. 123-135.

(24) Miller, S.L., and Orgel, L.E. 1974. The Origins of Life on the Earth. Englewood Cliffs, NJ: Prentice Hall.

(25) Miller, S.L.; Urey H.C.; and Oró, J. 1976. Origin of organic compounds on the primitive earth and in meteorites. J. Mol. Evol. 9: 59-72.

(26) Oparin, A.I. 1938. The Origin of Life. New York: Macmillan.

(27) Oró, J. 1960. Synthesis of adenine from ammonium cyanide. Biochim. Biophys. Res. Comm. 2: 407-412.

(28) Oró, J., and Kimball, A.P. 1961. Synthesis of purines under possible primitive earth conditions. I. Adenine from hydrogen cyanide. Arch. Biochem. Biophys. 94: 221-227.

(29) Oró, and Kimball, A.P. 1962. Synthesis of purines under possible primitive earth conditions. II. Purine intermediates from hydrogen cyanide. Arch. Biochem. Biophys. 96: 293-313.

(30) Reid, C., and Orgel, L.E. 1967. Synthesis of sugars in potentially prebiotic conditions. Nature 216: 455.

(31) Ring, D.; Wolman, Y.; Friedmann, N.; and Miller, S.L. 1972. Prebiotic synthesis of hydrophobic and protein amino acids. Proc. Natl. Acad. Sci. USA 69: 765-768.

(32) Sagan, C., and Khare, B.N. 1971. Long-wavelength ultraviolet photoproduction of amino acids in the primitive earth. Science 173: 417-420; also in Nature 232: 577-578.

(33) Sanchez, R.A.; Ferris, J.P.; and Orgel, L.E. 1966. Cyanoacetylene in prebiotic synthesis. Science 154: 784-785.

(34) Sanchez, R.A.; Ferris, J.P.; and Orgel, L.E. 1967. Synthesis of purine precursors and amino acids from aqueous hydrogen cyanide. J. Mol. Biol. 30: 223-253.

(35) Sanchez, R.A.; Ferris, J.P.; and Orgel, L.E. 1968. Conversion of 4-aminoimidazole-5-carbonitrile derivatives to purines. J. Mol. Biol. 38: 121-128.

(36) Schwartz, A.W., and Chittenden, G.J.F. 1977. Synthesis of uracil and thymine under simulated prebiotic conditions. Biosystems. 9: 87-92.

(37) Urey, H.C. 1952. On the early chemical history of the earth and the origin of life. Proc. Natl. Acad. Sci. USA 38: 363; also In The Planets. New Haven: Yale University Press, pp. 149-157.

(38) Van Trump, J.E., and Miller, S.L. 1972. Prebiotic synthesis of methionine. Science 178: 859-860.

(39) Van Trump, J.E., and Miller, S.L. 1980. The Strecker synthesis in the primitive ocean. In Proceedings of the 3rd ISSOL Meeting, ed. Y. Wolman, Jerusalem.

(40) Wolman, Y.; Haverland, W.H.; and Miller, S.L. 1972. Nonprotein amino acids from spark discharges and their comparison with the Murchison meteorite amino acids. Proc. Natl. Acad. Sci. USA 69: 809-811.

(41) Yoshino, D.; Hayatsu, R.; and Anders, E. 1971. Amino acids: catalytic synthesis. Geochim. Cosmochim. Acta 35: 927-938.

(42) Zeitman, B.; Chang, S.; and Lawless, J.G. 1974. Dicarboxylic acids from electric discharge. Nature 251: 42-43.

Mineral Deposits and the Evolution of the Biosphere, eds. H.D. Holland and M. Schidlowski, pp. 177-198. Dahlem Konferenzen, 1982.
Berlin, Heidelberg, New York: Springer-Verlag.

Stratified Sulfide Deposition in Modern and Ancient Environments

P. A. Trudinger* and N. Williams**
*Baas-Becking Geobiological Laboratory, Canberra City, A. C. T. 2601
**Carpentaria Exploration Co., Brisbane Q 4001, Australia

Abstract. Present day sulfide deposition occurs in many environments. One is at tectonically active sites, such as the East Pacific Rise and the Red Sea where sulfides, formed most likely by high-temperature abiotic sulfate reduction, and metals are supplied by hydrothermal fluids. Another is in nonthermal restricted environments such as the Black Sea where sulfides form by bacterial sulfate reduction, but where extreme sulfide accumulation is prevented by a lack of reactable metals.

Precambrian stratified sulfide deposits include sulfide-facies banded iron formations and black shales, volcanogenic massive sulfide deposits, shale-hosted Pb-Zn deposits, and sediment-hosted stratabound Cu deposits. Most show evidence of hydrothermal associations. There are, as yet, no unequivocal examples of Precambrian deposits formed entirely by biogenic sulfate reduction or of modern biogenic ore-forming environments.

INTRODUCTION

Precambrian stratified sulfide deposits are important sources of many metals, especially of Cu, Pb, and Zn. The stratification of minerals in the deposits is good evidence of sedimentary accumulation, although it can also develop in some instances below the sediment-water interface during diagenesis or by later selective replacement. Recognition of the low-temperature character of stratified sulfide depositional settings led to the hypothesis that the sulfides might, in some cases, have formed by microbial sulfate reduction, a process that is widespread in the modern environment. If it can be demonstrated that the hypothesis is correct, stratified sulfide deposits would provide clues to early developments in the biogeochemical cycle of sulfur and to the influence of these developments on Precambrian metallogeny.

In the present paper we discuss some of the main features of present-day and ancient sulfide deposits and draw some general conclusions regarding the possible contributions of biological processes to the development of Precambrian sulfide mineralization.

MODERN SULFIDIC ENVIRONMENTS

In modern environments sulfide can be derived from deep-seated mantle sources and by the abiological or biological reduction of sulfate. It is to the latter processes that this discussion is directed.

Abiological Sulfate Reduction

In modern environments dominated by hydrothermal activity, sulfate can be reduced abiologically by Fe^{2+} and by organic matter at deep crustal sites. In the laboratory such reactions only proceed at significant rates above 300°C (4,22,29). Reactions between hydrocarbons and sulfate are accelerated by the presence of sulfide; the sulfide may interact with sulfate to form elemental sulfur or polysulfides which then react with organic compounds to produce H_2S.

A proposed low temperature (80-120°C) reduction of sulfate by organic matter (26) has not been experimentally verified. For example, incubation of high specific activity ^{35}S-labelled sulfate for several weeks with a variety of crude petroleum fractions failed to produce radioactive sulfide at temperatures below 220°C, both in the presence and absence of H_2S (L.A. Chambers, P.A. Trudinger, and D.T. Rickard, unpublished results).

Biological Sulfate Reduction

The most important biological mechanism of sulfide production is dissimilatory sulfate reduction, a process in which sulfate replaces O_2 in the bacterial oxidation of organic matter (or H_2) under anaerobic conditions.

The free energy yields of these reactions are relatively low (about 10% of the free energy released in analogous O_2-linked oxidations) and large amounts of sulfate are reduced to sustain growth of the organisms. The "classical" dissimilatory sulfate-reducers, Desulfovibrio spp. and Desulfotomaculum spp., can carry out the incomplete oxidation of a few simple organic compounds (37). Recently, however, thanks largely to the work of F. Widdel, several genera have been described, among them organisms that catalyze the complete oxidation of higher fatty acids, and others which are capable of autotrophic growth on H_2, CO_2, and sulfate (N. Pfennig, personal communication).

Hydrothermal Deposits

The bulk of the constituents in hydrothermal deposits are derived from hydrothermal fluids; these are brines with concentrations of total dissolved solids ranging from about 3% to 50%. Compared with seawater, such brines are characterized by low pH, high SiO_2 content, high Ca/Mg ratios, and enrichment in Fe, Mn, and other metals (36).

Hydrothermal fluids may be derived from connate waters, by boiling during the crystallization of magmas, and by the deep circulation of seawater and meteoric water through the crust. The heated solutions are driven upwards either by thermal convection near hot igneous rocks or by processes such as compaction or syndepositional faulting. The solutions acquire their characteristic chemistry by interacting with the rocks through which they pass. Experiments on the interaction of seawater with basalt at high temperatures show that with time the original slightly basic Na-Mg-SO_4-Cl solution becomes an acidic, dominantly Na-Ca-Cl solution, containing significant quantities of Fe,

Mn, and other metals (4,22). Sulfate is depleted by the precipitation of anhydrite and, above 300°C, by reduction by Fe^{2+}.

Hydrothermal activity associated with oceanic spreading centers and rift zones is frequently accompanied by the deposition of metalliferous sediments (5). These may be oxidized (Mn- or Mn/Fe-rich) or sulfidic (Mn-poor); two hypotheses have been advanced to account for the differences between these types of sediments. The first suggests that sulfide-rich deposits form when hot, acidic hydrothermal fluids are exhaled directly onto the ocean floor, and that the precursors of oxidized deposits form when cooler fluids, modified beneath the surface by mixing with cool "ground waters" of approximately seawater composition, are exhaled (15). The second hypothesis proposes that exhalation of hydrothermal fluids into an environment of restricted circulation causes a sulfide-rich assemblage to be deposited while oxidized deposits arise from discharge of hydrothermal fluids into oxygenated, circulating seawater (5).

Environments with restricted circulation are presumably required for the preservation of hydrothermal sulfide deposits, but they are not essential to their formation. Sulfide deposits are forming today on the East Pacific Rise (EPR) near 21°N, 109°W (18) where sphalerite, pyrite, and chalcopyrite are precipitating around vents that discharge hot (∿350°C) fluids containing H_2S and metals. Based on preliminary data on the structural setting and sulfur isotopic composition of these deposits, Hekinian et al. (23) concluded that the sulfides had a magmatic source. However, these deposits, which are regarded by some as modern analogues of the Cyprus-type volcanogenic massive sulfide deposits, are unlikely to be preserved in the geological record. Their exposure to oxygenated waters causes oxidation of the sulfides, possibly catalyzed by the chemosynthetic, sulfide-oxidizing bacteria which have been detected in waters surrounding the vents.

The best characterized modern hydrothermal deposits are those of the Red Sea basin, located on an extension of the Indian mid-ocean ridge; these deposits consist of metalliferous sediments associated with waters of high salinity and elevated temperature in deeps near the central rift zone (12,33).

In the Atlantis II Deep, metals are concentrated to a much larger extent than in any other known modern marine sedimentary deposit, and distinct sulfide facies occur at depth (33).

Within the Atlantis II Deep an isolated pool of brine is trapped in a local depression of the Red Sea median valley about 2000 m below the surface of the sea. The brine exhibits thermal and salinity stratification; the pool is thermally active as evidenced by rises in the temperature of $\sim 4^{o}C$ between 1966 and 1971. The brines are anoxic (12a), but there is little evidence of highly reducing conditions: redox potentials of interstitial waters in the sediments are of $\sim$+300 to +400 mV down to depths of 850 cm (12b).

The sediments of the Atlantis II Deep are well bedded and partly laminated. The sulfide facies consist of black, fine-grained sediment and appear to spread continuously throughout the deep. Pyrite, Fe-monosulfide, chalcopyrite, and sphalerite are the main metallic minerals of the sulfide facies and are ubiquitous minor components of the uppermost amorphous-silicate zone. The lowermost sulfide facies is thought to have been deposited immediately after continuous geothermal activity began about 15,000 y ago (33).

The solids of the Atlantis II sediments were probably derived from the overlying metalliferous brines which formed by the interaction of Red Sea water with underlying Miocene evaporites. The origin(s) of sulfide-sulfur in these deposits is uncertain. In an underlying detrital-oxidic-pyritic zone, where $\delta^{34}S$ values range from -33.8 to $\sim$+10‰, two sources of sulfur seem possible (33). In general, ^{32}S enrichment is associated with detrital phases which are rich in pyrite and organic matter. It seems likely that the ^{32}S-enriched sulfides were formed from biogenic sulfur. On the other hand, ^{34}S enrichment is characteristic of samples containing significant amounts of chalcopyrite or sphalerite which appear to be derived from hydrothermal sources.

Sulfides in the two sulfide facies and in the presently depositing amorphous-silicate zone are more uniform isotopically; (sulfide versus Red Sea sulfate) are 1.010 to 1.017. While these factors are within the range found during bacterial sulfate reduction (10), they are also close to those of sulfate-sulfide pairs from known hydrothermal deposits and are lower than those frequently observed in other marine sediments where sulfide is clearly of biogenic origin. Moreover, the brines and sediments of the Atlantis II deep appear to be sterile, although bacterial sulfate reduction has been detected in the transition zone between brine and seawater. Perhaps some evidence in favor of a non-biogenic origin for the Atlantis II sulfides is to be found in the fact that fractionation factors characteristic of biogenic sulfides (on the order of 1.050) are observed for sulfate-sulfide (mainly pyrite) pairs in the nearby Discovery Deep and in Red Sea sediments away from hydrothermal regions (12c).

Biogenic Sulfides

Sulfate-reducing bacteria are strict anaerobes; as a group they are adaptable to almost any anoxic environment on Earth. Representatives of the group, or evidence of their presence, can be found in soils, in fresh, marine and brackish waters, in artesian waters, in hot springs, in geothermal areas, in Antarctic lakes, in deep sea sediments, in estuarine muds, and in oil and natural gas wells. In sedimentary environments sulfate may be reduced in the water column. However, most sulfate reduction takes place close to the sediment/water interface; below this interface rates of sulfate reduction decline exponentially with depth due to the depletion of degradable organic matter and perhaps due to the lack of other nutrients. Nevertheless, S-isotope profiles in some deep sea sediments have been interpreted as indicating the reduction of sulfate at depths of several hundred meters during time periods of several million years (20).

Fe-sulfides are by far the most common and abundant metal sulfides formed biogenically in modern environments. Berner (2) outlined three principal factors limiting the amount of Fe-sulfide which may form in a sediment: (a) the availability of sulfate, (b) the concentration and reactivity of Fe-compounds, and (c) the concentration of organic matter which can be utilized by sulfate-reducing bacteria. These are discussed in turn below:

(a) Sulfate may be limiting in lacustrine environments and in organic-rich, sub-surface marine sediments where the rate of sulfate reduction may exceed the rate of sulfate replenishment. Sulfate is undetectable, for example, in the bottom waters of Ace Lake, Antarctica, and in the interstitial waters of a number of marine sediments. However, because the main zone of sulfate reduction in sediments is close to the sediment-water interface where diffusion from the overlying waters is relatively effective, the sulfide content of sediments in saline environments is rarely limited by sulfate supply. More commonly, an "excess" of sulfur is observed, i.e., the total reduced sulfur exceeds that expected from the amount of sulfate originally buried with the sediment (19).

(b) The main sources of Fe in sediments are detrital Fe-minerals which vary in their reactivity with H_2S. In particular, sand and silt-sized grains of magnetite and ferruginous silicates are relatively unreactive and may be preserved as metastable phases in basically pyritic sediments. It is noteworthy, however, that in Black Sea sediments up to 40% of potentially reactive Fe (excluding silty material) has not in fact reacted with H_2S (13), although the latter does not appear to be limiting in this environment. We have also found (P.A. Trudinger, L.A. Chambers, and P.J. Cook, unpublished results) that nonsulfidic Fe in some anoxic, sulfidic, deep sea sediments can react stoichiometrically and instantaneously with H_2S in the laboratory. The question of reactivity of Fe with H_2S in sediments is one that merits further study.

(c) Probably the most important factor governing biogenic H_2S production is the supply of organic matter. Certainly, rates of sulfate reduction in sediments and waters can be markedly stimulated by the addition of readily metabolizeable organic compounds. Moreover, rates of reduction in a variety of sediments correlate linearly with rates of sedimentation, a fact that accords with the observation that total organic carbon preserved in sediments tends to be positively correlated with sedimentation rate (20). As a consequence, there is a general linear correlation between the organic carbon and the reduced sulfur content of marine sediments. It is important to note that this correlation is not necessarily related to the rate of organic matter production in marine settings since the extent to which sulfate reduction contributes to the overall degradation of organic matter varies widely. In the Black Sea, for example, most of the organic matter produced by phytoplankton is consumed in the upper oxic waters, and only about 5% is available to support sulfate reduction. In near shore, shallow water environments, however, 50% or more of primary organic carbon may be degraded by sulfate-reducing bacteria.

A further factor governing the preservation of sulfides is the long-term maintenance of anoxic conditions. Organic matter is important here as well, since it is the oxidation of organic matter by aerobic microorganisms which is the main cause of oxygen depletion in the modern aquatic environments. Marine sediments are often anoxic at depth because diffusion of oxygen from the overlying water is quite slow. The most likely environments for significant biogenic sulfide accumulation, however, are those in which reducing conditions become established above the sediment water interface. These conditions can be found in many lakes, silled basins, and local depressions in the open ocean, particularly where density stratification, caused by steep thermal and/or salinity gradients, minimizes exchange between upper and lower water layers.

The largest and best studied anoxic basin is the Black Sea (13), a silled basin with a maximum depth of about 2000 m and with a well defined halo-, thermocline at 150-200 m depth. Below this depth anoxic conditions prevail and the H_2S concentration increases to an average of about 10 mg 1^{-1}. Analysis of sediment cores indicates that the Black Sea was originally a well-aerated fresh water lake which, beginning about 9000 y B.P., has been gradually transformed to its present state by the influx of Mediterranean seawater via the Bosporus following climatic warming and the retreat of the ice. The present environmental conditions were established about 3000 y ago. The surface muds of the Black Sea are sites of intensive bacterial sulfate reduction, and sulfide there has a typical biogenic isotopic signature (∿ - 50‰ relative to Black Sea sulfate). A second site of bacterial sulfate reduction is located just below the H_2S-O_2 transition zone.

Iron sulfides (including framboidal pyrite) are present throughout the sediments with the greatest concentration (up to 3% S) in an organic-rich sapropel unit located about 30 cm below the sediment surface. In sediments with a high organic content (>2%C) the bulk of the sulfides are formed in the sediments. Where the organic carbon is low, however, most of the sulfides present in the sediments formed during sedimentation through the H_2S-laden overlying water. Less than 10% of biologically produced sulfide, however, is fixed: the remainder is recycled at the O_2-H_2S transition zone.

Tidal Algal Flats

Stratigraphic and mineralogical similarities have been noted between a number of stratiform deposits (e.g., copper deposits in the southwestern USA; see below) and coastal sabkhas - the evaporite flats that form along subaerial landward margins of regressive seas (34). Sabkhas are bordered on the landward side by extensive cyanobacterial mats and a proposed model for metal mineralization involves: (a) burial of the algal mat; (b) development of bacterial sulfate reduction within the buried mat; and (c) fixation of metal as a constituent of a sulfide, from metal-enriched continental groundwater which passes through the buried mat as the result of evaporative pumping.

No sound evidence for this model is yet forthcoming from modern environment studies. While H_2S has been detected in sabkha sediments of the Trucial Coast, Arabian Gulf, the presence and preservation of metal sulfides has not been reported. Tidal flats may not, in fact, be ideal environments for sulfide preservation. Quantitative studies of primary organic productivity, sulfate reduction, and sulfide stability in evaporitic intertidal algal mat sediments in the northeastern region of Spencer Gulf, South Australia, indicate that the organic carbon and iron sulfides are largely recycled within a few months (J. Bauld, L.A. Chambers, and G.W. Skyring, unpublished results). More favorable shallow water marine environments for sulfide preservation may be found in coastal lagoons, such as the Coorong Lagoon in South Australia and Solar Lake, Sinai, where considerable thicknesses of organic matter have accumulated.

ANCIENT SULFIDIC SEDIMENTARY ROCKS

There are many different kinds of sulfidic sedimentary rocks in the Precambrian. For the purposes of this review, we have divided them into five groups: (a) sulfide-facies banded iron formations (BIF's), (b) volcanogenic massive sulfide deposits, (c) shale-hosted stratiform Pb-Zn deposits, (d) sediment-hosted stratabound Cu deposits, and (e) black shales. While these groups contain fairly distinct lithologies, the boundaries between the groups are by no means sharply defined, and differences between the groups could well be gradational. Bearing in mind this shortcoming in our classification, we consider in the following discussion the particular features of the several groups and examine the environments in which they appear to have formed.

Sulfide-Facies BIF's

This group consists of those BIF's in which Fe occurs largely as a constituent of sulfides (25). In the Archean (>2.6 Gy), sulfide-facies BIF's occur with carbonate-, oxide-, and silicate-facies BIF's in the curvilinear volcano-sedimentary greenstone belts within regions of granites and gneisses. The Michipicoten region of Canada is a classic example of such a belt. They occur in environments dominated by shaly sediments and the younger pyroclastic phases of tholeiitic to calc-alkaline volcanism. Sulfide-facies BIF's grade into carbonate-facies BIF's, and these, in turn, grade into oxide facies BIF's. The transition is accompanied by a decrease in volcanics and an increase in conglomeratic sediments, suggesting that the sulfide facies formed in a deep water environment dominated by volcanism. The sulfide facies consists of massive sulfides (up to 38% S), is carbonaceous, and is enriched in As, Mn, Zn, Cu, Ni, and Au (21). The major sulfides are pyrite and pyrrhotite. Similar sulfide-facies BIF's are widespread in southern Africa and Western Australia (16). The oldest are in the 3.7 Gy old Isua sequence in Greenland where, in contrast to the younger examples, the main sulfides are chalcopyrite and pyrrhotite (24).

S-isotopic ratios in most sulfide-facies BIF's have a mean value close to 0‰ and a small spread around the mean (32). This suggests that the sulfide-S is magmatic, and that there is no significant biogenic component. However, the youngest examples (∿2.75 Gy old), from Woman River and Michipicoten, have different distributions. The mean value of $\delta^{34}S$ is close to 0‰, but $\delta^{34}S$ values extend up to 10‰ on both sides of the mean (21). These distributions are regarded by some as the oldest known evidence for bacterial sulfate reduction (21,32).

The origin of the Archean sulfide-facies BIF's is difficult to determine because exposures are poor, and later metamorphism and deformation have been intense. Because they occur in restricted sedimentary environments dominated by volcanism, are associated with fine-grained carbonaceous and tuffaceous sediments, and have magmatic $\delta^{34}S$ values and high contents of various base metals and Au, their origin is widely considered to be be volcanic-exhalative, in which hydrothermal fluids containing Fe and other metals were exhaled into relatively deep water, and where the banded sulfidic sediments were formed as chemical precipitates when these fluids mixed with seawater. The presence of abundant carbonaceous matter may simply indicate the presence of quiet deep-water anoxic environments, or that the sites of volcanic exhalation were particularly favorable for the growth of organisms, for instance, the bacteria that utilize the oxidation of reduced sulfur (8).

Sulfide-facies BIF's are also widespread in the Proterozoic. In the Lower Proterozoic (∿2.6 Gy to ∿1.9 Gy B.P.), the period of formation of the great Superior-type BIF's, there was also a restricted formation of sulfide-facies BIF's. The classic example is the Wauseca Pyritic Member of the Iron River-Crystal Falls district of Michigan (24). The Member is pyritic and carbonaceous (up to 29.0% S and 18.2% C) and is thought to have formed in an offshore, deep-water anoxic environment. The sulfides could have formed by bacterial sulfate reduction, but there are no confirming S-isotopic data. The source of Fe in Superior-type BIF's is problematic and beyond the scope of this review. An often postulated source is volcanism, but the geological links between the Wauseca Pyritic Member and volcanism are not obvious, as they are with Archean sulfide-facies BIF's.

Younger Proterozoic sulfide-facies BIF's are also known. They are geographically restricted in distribution and are closely associated with stratiform base-metal sulfide deposits in sediments, rather than with other BIF facies. Good examples of such deposits are the highly pyritic dolomitic siltstones adjacent to the 1.7 Gy old north Australian shale-hosted stratiform Pb-Zn deposits at McArthur River, Mount Isa, and Hilton (27), and the pyritic siltstones and argillites hosting the >1.43 Gy old Sullivan deposit in British Columbia (9). These BIF's are almost certainly related to the associated Pb-Zn deposits and their origin is discussed later.

Volcanogenic Massive Sulfide Deposits

These deposits occur in rocks of all ages and are an economically important type of stratified sulfide deposit containing significant resources of Cu, Pb, and Zn. In this review we are concerned with the Precambrian examples, and these are best known from the younger Archean greenstone belts of Canada. They are dominantly Cu-Zn deposits, and are typically associated with intermediate to acidic volcanics. Their features have been widely reviewed (31).

In simplest terms, the deposits consist of three parts, a lower funnel-shaped (widest part to the top) discordant zone of Cu-rich stringer ore with an accompanying envelope of hydrothermal alteration, a central more stratiform zone of massive Zn-rich sulfide ore, and, in some instances, an upper blanket-shaped "exhalite" zone of chert. Usually the deposits have undergone post-mineralization deformation and metamorphism, and it is difficult to recognize primary depositional features. The deposits are widely regarded as having been formed by submarine volcanic-exhalative processes. These have been modeled by Large (26), who concluded that mineralization occurred at the top of volcanic piles, just below and on the sea floor, and that precipitation of sulfides followed mixing of seawater with hot, mildly acid, reduced chloride-rich fluids. Little is known about the water depths at which mixing occurred. $\delta^{34}S$ (sulfide) values in the deposits cluster around 0‰, indicating that the S has a magmatic rather than a biogenic origin (26).

There are some interesting differences between Archean and Phanerozoic volcanogenic massive sulfide deposits that may reflect a change through time from reducing to oxidizing conditions in seawater (26). These include: (a) a lack of sulfates in the upper exhalative zones of the Archean deposits, in contrast to the Phanerozoic deposits, (b) a lack of change in $\delta^{34}S$ (sulfide) values up through Archean deposits, in contrast to Phanerozoic deposits in which the $\delta^{34}S$ values tend to decrease upwards suggesting increasingly oxidizing conditions towards the sea floor, and (c) a prevalence of pyrite in exhalite layers overlying the Archean deposits, in contrast to a prevalence of hematite in the exhalites overlying Phanerozoic deposits. It is critical that these three aspects be documented in deposits of Proterozoic age, as this information could well provide important new insights into the timing of the change-over from reducing to generally oxidizing conditions in the oceans.

Shale-hosted Stratiform Pb-Zn Deposits

These include some of the world's largest Pb-Zn deposits. The best known examples are the previously mentioned north Australian deposits at McArthur River, Mount Isa, and Hilton, and the Sullivan deposit in British Columbia. All these deposits are Middle Proterozoic in age. However, Phanerozoic examples are also known; the most significant of these are the recently discovered Silurian deposits at Howards Pass, Canada.

The McArthur River deposits are the least metamorphosed and best preserved Proterozoic examples of this type of deposit. They occur in carbonaceous and pyritic siltstones of the Barney Creek Formation and consist of thin layers of sulfides interbanded with sulfide-poor dolomitic siltstones; pyrite, the most abundant sulfide, occurs as tiny (average diameter $<2\mu m$) crystals arranged in laminae and framboids. Framboidal pyrite is a common pyrite habit in modern biogenous sediments and has often been taken as evidence of biogenic sulfide formation. It should be noted, however, that framboids can be produced readily in the laboratory without benefit of biological activity (19). Sphalerite and galena occur interstitially to the pyrite in diffuse laminae that are parallel to the bedding (27,38).

The host siltstones occur in the midst of a thick sequence of shallow-water stromatolitic and evaporitic dolosiltstones and related sediments that formed in coastal sabkha and/or saline lacustrine environments (38). These siltstones, apparently represent a slightly deeper depositional environment than that of the adjacent sabkha sediment. They were deposited in a half-graben formed by movement on an adjacent syndepositional fault, the Emu Fault. In the Emu Fault zone there are small discordant deposits of Cu, Pb, and Zn that resemble Mississippi Valley-type deposits. The geology and geochemistry of these deposits indicate that they formed along channelways for the hydrothermal fluids that formed the stratiform deposits and that the fluids came from the direction of the Emu Fault (38). It is thought that the McArthur deposits were deposited from either volcanically-derived mineralizing fluids or metalliferous basinal brines, and that sulfide deposition occurred either during exhalation of these fluids (27) or during their introduction into the host sediments soon after deposition (38). S-isotopic data support a hydrothermal origin for sulfur in galena and sphalerite, and a different and possibly bacterial origin for pyritic sulfur (35). The sulfide-facies BIF's mentioned previously are restricted to regions adjacent to Pb-Zn mineralization, suggesting that the BIF's are related to the same hydrothermal event that produced the mineralization. Remote from the Pb-Zn mineralization there is possible evidence, in the less pyritic host sediments, that this pyrite formed by bacterial sulfate reduction. The evidence includes the laminar and framboidal habit of the pyrite and a strong positive correlation between the S and C content of the sediments (38). The other north Australian deposits are very similar to the McArthur deposits and are thought to have had a similar origin.

At Sullivan much of the ore is very similar to that in the north Australian deposits. In the center of the deposit, however, there is a zone of massive pyrite, pyrrhotite, and chlorite, and beneath the deposit there is a pipe of intense hydrothermal alteration reminiscent of the alteration pipes beneath volcanogenic massive sulfide deposits. The footwall at Sullivan is brecciated and partly conglomeratic; these features are considered to indicate that the deposit is centered on a depression thought to be caused by collapse associated with the same hydrothermal activity that led to ore formation by exhalative processes. There are no data available to indicate water depths during exhalation. However, no shallow-water features have ever been recorded. S-isotopic data do not provide strong support for bacterial sulfate reduction (9).

Sediment-hosted Stratabound Cu Deposits

These deposits become prominent in Late Proterozoic sediments, but they are also found in younger Phanerozoic sediments. The best documented Proterozoic deposits include the famous Copperbelt deposits of Zambia and Zaire (1,17) and the White Pine deposit in Michigan (6). The classic Phanerozoic example is the Kupferschiefer of Europe. In contrast to the previously described stratified base-metal sulfide deposits, the Cu deposits are not massive. The concentration of disseminated sulfides in such deposits is typically low (<10 wt %S).

The >1.05 Gy old White Pine deposit has undergone little post-depositional metamorphism, and primary depositional features are well preserved. The deposit is hosted by the carbonaceous and pyritic Nonesuch Shale, which is a shallow water unit that underwent occasional emergence and dessication. The mineralization consists of disseminated chalcocite at the base of the Nonesuch Shale, immediately above the Copper Harbor Conglomerate.

In detail, the ore zone appears strictly conformable to the thin layering of the host siltstone, but careful regional mapping has shown that it transects bedding at a very low angle. Between the pyritic hanging wall sediments and the copper ore there is a transitional zone of Cu-Fe sulfides that partially replace pyrite. The deposit is thought to have formed in two stages, the first involving the formation of disseminated and framboidal pyrite by bacterial sulfate reduction during early diagenesis, the second involving the introduction of Cu-bearing fluids and the formation of Cu sulfides by pyrite replacement. Strong evidence for bacterial sulfate reduction includes the isotopic composition of the sulfides (7), their framboidal habit, and a strong positive correlation between S and C in the sediments (38).

The origin of the Copperbelt deposits is less well understood, because later deformation and metamorphism has obscured many of the initial features. Some deposits, such as the Zairean Kamoto deposit, resemble the White Pine deposit and probably had a similar origin (1). The Zambian deposits occur in both sandstones and shales of the Roan Group that were deposited in very shallow water in quiet and evaporitic (sabkha), epeiric marine or lacustrine conditions (11). Metamorphism has obscured critical evidence needed to determine how Cu was deposited. Possible explanations are precipitation either during or after sedimentation by microbial sulfate reduction, by two-stage pyrite replacement processes, or by inorganic sulfate reduction.

Black Shales

In the Precambrian, black shales containing only minor iron sulfides are far more widespread than any of the above stratified sulfide deposits. They are much more representative of ordinary sulfidic sediments formed in normal environments than any of the above deposit types, most of which apparently formed in unusual environments perturbed by relatively rare processes such as submarine volcanism and the injection of hydrothermal fluids.

Black shales have received little attention to date; this is unfortunate, as they are an important key to the evolution of the C and S cycles. The best documented Archean examples are in Canada (8). They are typically tuffaceous; like the more sulfidic Archean deposits, they are carbonaceous, enriched in Cu, Zn, As, Ag, Sn, Sb, Hg, and Pb, and have sulfide $\delta^{34}S$ values of $\sim 0‰$.

Some chemical data are also available for Proterozoic black shales, including the Canadian Aphebian (2.5 to 1.6 Gy B.P.) shales analyzed by Cameron and Garrels (8), the 1.7 Gy old Australian Barney Creek Formation remote from the McArthur mineralization (38), and the >1.05 Gy old Nonesuch Shale, Michigan, remote from the White Pine mineralization (6). In these shales $\delta^{34}S$ values of sulfides are typically greater than 0‰; this evidence, together with the framboidal and laminar pyrite habits and the good correlation between S and C content of the sediments (38), implies that by Proterozoic times bacterial sulfate reduction was an important process in the formation of sulfides under anoxic conditions.

CONCLUDING REMARKS

At the present time there are no known Precambrian sulfide deposits which can, with any certainty, be identified as biogenic. There are also no known modern biogenic sulfide deposits of any quantitative significance. Nevertheless, when we compare various examples of ancient stratified deposits with those forming in biologically active environments today, some similarities do emerge, particularly in the case of those ancient deposits containing only minor sulfides. For example:

1. There is a positive correlation between reduced sulfur and organic carbon concentrations.

2. Framboidal pyrite, a frequent pyrite habit in modern sediments, is also found in some ancient deposits, especially where a biogenic component seems most plausible, e.g., in the north Australian Pb-Zn deposits and in the stratabound Cu deposits.

3. Various S-isotopic patterns recognized in modern sediments have their counterparts in ancient environments.

Perhaps these similarities indicate that processes involved in sulfide deposition today were also operating in the Precambrian.

Large-scale sulfide deposition, both ancient and modern, appears to be restricted to environments with clear-cut hydrothermal associations. Indeed, the abiogenic EPR deposits and those of the Red Sea may be modern analogues of volcanogenic massive sulfides and possibly of some shale-hosted Pb-Zn deposits (e.g., Sullivan).

In biologically active environments like the Black Sea, the main factor limiting sulfide accumulation is almost certainly the lack of metals, since rates of sulfide formation appear to be adequate for large-scale sulfide deposition (37). On the other hand, the metalliferous, hydrothermal Atlantis II Deep lacks biological activity. It is perhaps to environments intermediate between these two extremes that we should look for biogenic sulfide accumulation. Possible candidates are the anoxic lakes of the East African Rift System which are supplied with metals from hydrothermal sources and which support a vigorous biological population (14).

Acknowledgements. The Baas Becking Laboratory is supported by the Australian Mineral Industries Research Association Ltd., the Bureau of Mineral Resources, and the Commonwealth Scientific and Industrial Research Organization.

REFERENCES

(1) Bartholomé, P., ed. 1974. Gisements stratiformes et provinces cuprifères, pp. 203-214. Liège: Société Géologique de Belgique.

(2) Berner, R.A. 1971. Principles of Chemical Sedimentology. New York: McGraw-Hill.

(3) Berner, R.A. 1978. Sulfate reduction and the rate of deposition of marine sediments. Earth Planet. Sci. Lett. 37: 492-498.

(4) Bischoff, J.L., and Dickson, F.W. 1975. Seawater-basalt interaction at 200° and 500 bars: implications for origin of sea-floor heavy-metal deposits and regulation of seawater chemistry. Earth Planet. Sci. Lett. 25: 385-397.

(5) Bonatti, E. 1975. Metallogenesis at oceanic spreading centers. Ann. Rev. Earth Planet. Sci. 3: 401-431.

(6) Brown, A.C. 1971. Zoning in the White Pine copper deposit, Ontonogan County, Michigan. Econ. Geol. 66: 543-573.

(7) Burnie, S.W.; Schwarcz, H.P.; and Crocket, J.H. 1972. A sulfur isotopic study of the White Pine mine, Michigan. Econ. Geol. 67: 895-914.

(8) Cameron, E.M., and Garrels, R.M. 1980. Geochemical compositions of some Precambrian shales from the Canadian Shield. Chem. Geol. 28: 181-197.

(9) Campbell, F.A.; Ethier, V.G.; Krouse, H.R.; and Both, R.A. 1978. Isotopic compositions of sulfur in the Sullivan orebody, British Columbia. Econ. Geol. 73: 246-268.

(10) Chambers, L.A., and Trudinger, P.A. 1979. Microbiological fractionation of stable sulfur isotopes: a critical review. Geomicrobiol. J. 1: 249-293.

(11) Clemmey, H. 1978. A Proterozoic lacustrine interlude from the Zambian Copperbelt. Spec. Publs. Int. Ass. Sediment. 2: 257-278.

(12) Degens, E.T., and Ross, D.A., eds. 1969. Hot Brines and Recent Heavy Metal Deposits in the Red Sea, (a) pp. 138-147; (b) pp. 190-193; (c) pp. 474-498. New York: Springer-Verlag.

(13) Degens, E.T., and Ross, D.A., eds. 1974. The Black Sea. Geology, Chemistry and Biology, p. 530. Tulsa: American Association of Petroleum Geologists.

(14) Degens, E.G.; von Herzen, R.P.; Wong, H.-K.; Deuser, W.G.; and Jannasch, H.W. 1973. Lake Kivu: Structure chemistry and biology of an east African rift lake. Geol. Rund. 62: 245-277.

(15) Edmond, J.M. et al. 1979. On the formation of metal-rich deposits at ridge crests. Earth Planet. Sci. 46: 19-30.

(16) Eichler, J. 1976. Origin of the Precambrian banded iron-formations. In Handbook of Strata-bound and Stratiform Ore Deposits, ed. K.H. Wolf, vol. 7, pp. 157-201. Amsterdam: Elsevier.

(17) Fleischer, V.D.; Garlick, W.G.; and Haldane, R. 1976. Geology of the Zambian Copperbelt. In Handbook of Strata-bound and Stratiform Ore Deposits, ed. K.H. Wolf, vol. 6, pp. 223-352. Amsterdam: Elsevier.

(18) Franchetau, J. et al. (CYAMEX Scientific Team) 1979. Massive deep-sea sulphide ore deposits discovered on the East Pacific Rise. Nature 277: 523-528.

(19) Goldhaber, M.B., and Kaplan, I.R. 1974. The sulfur cycle. In The Sea, ed. E.D. Goldberg, vol. 5, pp. 569-655. New York: John Wiley and Sons.

(20) Goldhaber, M.B., and Kaplan, I.R. 1975. Controls and consequences of sulfate reduction rates in recent marine sediments. Soil. Sci. 119: 42-55.

(21) Goodwin, A.; Monster, J.; and Thode, H.G. 1976. Carbon and sulfur isotope abundances in Archean iron-formations and Early Precambrian life. Econ. Geol. 71: 870-891.

(22) Hajash, A. 1975. Hydrothermal processes along mid-ocean ridges: an experimental investigation. Contrib. Mineral. Petrol. 53: 205-226.

(23) Hekinian, R.; Fevrier, M.; Bischoff, J.L.; Picot, P.; and Shanks, W.C. 1980. Sulfide deposits from the East Pacific Rise near 21°N. Science 207: 1433-1444.

(24) James, H.L.; Dutton, C.E.; Pettijohn, F.J.; and Wier, K.L. 1968. Geology and ore deposits of the Iron River - Crystal Falls district, Iron County Michigan, U.S. Geol. Surv. Prof. Pap. 570.

(25) James, H.L., and Sims, P.K. 1973. Precambrian iron-formations of the world. Econ. Geol. 68: 913-914.

(26) Large, R.R. 1977. Chemical evolution and zonation of massive sulfide deposits in volcanic terrains. Econ. Geol. 72: 549-573.

(27) Lambert, I.B. 1976. The McArthur zinc-lead-silver deposit: features, metallogenesis and comparisons with some other stratiform ores. In Handbook of Strata-bound and Stratiform Ore Deposits, ed. K.H. Wolf, vol. 6, pp. 535-585. Amsterdam: Elsevier.

(28) Monster, J.; Appel, P.W.U.; Thode, H.G.; Schidlowski, M.; Carmichael, C.M.; and Bridgwater, D. 1979. Sulfur isotope studies in early Archean sediments from Isua, West Greenland: implications for the antiquity of bacterial sulfate reduction. Geochim. Cosmochim. Acta 43: 405-413.

(29) Ohmoto, H., and Rye, R.O. 1979. Isotopes of sulfur and carbon. In Geochemistry of Hydrothermal Ore Deposits, ed. H.L. Barnes, pp. 509-567. New York: John Wiley and Sons.

(30) Orr, W.L. 1974. Changes in sulfur content and isotopic ratios of sulfur during petroleum maturation - Study of Big Horn Basin Paleozoic Oils. Bull. Am. Assoc. Petrol. Geol. 58: 2295-2318.

(31) Sangster, D.F., and Scott, S.D. 1976. Precambrian, strata-bound, massive Cu-Zn-Pb sulfide ores of North America. In Handbook of Strata-bound and Stratiform Ore Deposits, ed. K.H. Wolf, vol. 6, pp. 129-222. Amsterdam: Elsevier.

(32) Schidlowski, M. 1979. Antiquity and evolutionary status of bacterial sulfate reduction: sulfur isotope evidence. Origins of Life 9: 299-311.

(33) Shanks, W.C., and Bischoff, J.L. 1980. Geochemistry, sulfur isotope composition, and accumulation rates of Red Sea geothermal deposits. Econ. Geol. 75: 445-449.

(34) Smith, G.E. 1976. Sabkha and tidal-flat facies control of stratiform copper deposits in North Texas. In Handbook of Strata-bound and Stratiform Ore Deposits, ed. K.H. Wolf, vol. 6, pp. 407-446. Amsterdam: Elsevier.

(36) Tooms, J.S. 1970. Review of knowledge of metalliferous brines and related deposits. Trans. Inst. Min. Metall. 79: B116-B126.

(37) Trudinger, P.A. 1976. Microbiological processes in relation to ore genesis. In Handbook of Strata-bound and Stratiform Ore Deposits, ed. K.H. Wolf, vol. 2, pp. 135-190. Amsterdam: Elsevier.

(38) Williams, N. 1978. Studies of the base-metal sulfide deposits at McArthur River, Northern Territory, Australia: I and II. Econ. Geol. 73: 1005-1056.

Mineral Deposits and the Evolution of the Biosphere, eds. H.D. Holland and M. Schidlowski, pp. 199-218. Dahlem Konferenzen, 1982.
Berlin, Heidelberg, New York: Springer-Verlag.

Banded Iron Formation: Distribution in Time and Paleoenvironmental Significance

H. L. James* and A. F. Trendall**
*1617 Washington St., Port Townsend, WA 98368, USA
**Geological Survey of Western Australia, Perth, W. A. 6000, Australia

Abstract. Iron-rich sediments have been deposited intermittently throughout earth history, but virtually all significant deposits of the cherty layered rock referred to as iron formation are of Precambrian age. Present evidence, by no means conclusive, suggests three periods of peak deposition rate: mid-Archean (age 3400-2900 m.y.), early Proterozoic (age 2500-1900 m.y.), and late Proterozoic (age 750-500 m.y.). The deposits of early Proterozoic age far outweigh those of all other ages; in the aggregate they account for 90 percent or more of the estimated total of 10^{15} tons of originally deposited iron formation.

This review of the environmental significance of banded iron formations deals only with those of the cherty oxide facies. Two distinct subtypes are recognized: well laminated cherty banded iron formation (BIF in this paper) and granule iron formation. Together these form by far the greatest bulk of all rocks collectively called iron formation. Both are primary chemical precipitates modified subsequently by various processes which have not changed the initial chemical composition of the sediments significantly. From geological evidence, the most clearly established paleoenvironmental characteristics of the depositional basins of BIF were unusual tectonic and environmental stability coupled with an absence of clastic contamination; a desert environment is possible but not proven. There is insufficient evidence to demand a direct volcanic source for any of the precipitated components of BIF, and plausible models involving continuing local precipitation from ocean water circulating through a barred basin are consistent with geological evidence; so also is a climatically coupled mechanism of precipitation, perhaps a biochemical mechanism. While it is clear that the paleoenvironment of granule iron formation was, by contrast, a high-energy one in which the deposited material was subjected to para-depositional reworking, it is not clear to what extent the original materials were deposited by a mechanism like that of the BIFs. Granule iron formations are texturally similar to many modern and ancient limestones, but it does not follow that they are replaced carbonates.

DEFINITION, CLASSIFICATION, AND NOMENCLATURE

Banded iron formation is a thinly layered sedimentary rock consisting mainly of silica and iron minerals; it is believed to be an original chemical precipitate modified by diagenesis and metamorphism. Most deposits are of Precambrian age. The iron content is generally in the range of 24-35%, representing a 5- to 7-fold increase over normal crustal abundance. Not included in this review are the non-cherty ironstones of Phanerozoic age, which, though of common occurrence and locally of considerable economic value, are of somewhat different origin and in any case represent much smaller concentrations of iron.

The first attempt to formulate an accurate definition of "iron formation" effectively embraced all sedimentary rocks containing 15% or more Fe and proposed a classification into four facies: oxide, carbonate, silicate, and sulfide (20). These facies were thought to reflect not merely distinctive compositional varieties, but also to represent environmentally controlled variants within the depositional continuum of a single barred basin.

Since 1954 various workers have proposed alternative schemes of classification and nomenclature, and international unanimity has yet to be achieved (see (4,24)). For the present purpose the name "iron formation" is unfortunate insofar as it may imply a unique paleoenvironmental setting and a genetic significance which may not exist; by analogy, "calcium formation" would not be helpful to the extent that it suggested a common origin and significance for all limestones, as well as for sedimentary anhydrite and gypsum.

ABUNDANCE AND AREAL DISTRIBUTION (H.L. James)

Occurrences of iron formation probably number in the thousands. Most beds are thin, from less than a meter to a few meters, and of limited extent, but some are hundreds of meters thick and are preserved over tens of thousands of square kilometers. Several partial listings of occurrences have been published (5,15,21,24), but no complete tabulation is available. The list of selected deposits presented in Table 1 obviously is far from exhaustive, but it does include virtually all of the world's major deposits and nearly all of those deposits for which reasonable assessments of age and initial size can be made.

Estimates of initial size, expressed qualitatively in Table 1, are presented graphically in Figure 1. The reliability of the numerical appraisals varies widely. For deposits such as those of the Hamersley Range of Western Australia that have not undergone extensive post-depositional deformation and erosion, the estimate may be good to a factor of about two. But for the more typical deposits, for which data on thickness and distribution are scattered and incomplete, estimates have at best only an order-of-magnitude validity.

Large deposits of iron formation are present in all major Precambrian cratons of the earth, and probably 90% or more of all iron formation that has been preserved is contained in the five great districts placed in the "Very large" class in Table 1: the Labrador Trough and its extensions in Canada; the Hamersley Range of Western Australia; the Quadrilatero Ferrifero of Minas Gerais, Brazil; the Transvaal-Griquatown Basins of South Africa; and the Krivoy Rog-KMA areas of the USSR. The total initial magnitude represented by these deposits is about 10^{15} tonnes of iron formation, containing about 30% Fe. The possibility that comparable deposits are yet to be located is slight - the ease of recognition and the economic incentive for discovery argue against it - except perhaps in almost totally inaccessible sites, such as beneath the great ice caps of Greenland and Antarctica. The possibility that some major deposits were formed and later stripped completely away cannot be completely dismissed but also seems unlikely; even in deeply eroded terrains some remnants are likely to survive through vagaries of

TABLE 1 - Estimated intial size and age of selected deposits of cherty banded iron formation.

	Ref. No.[1]	Area (May include more than one stratigraphic unit)	Class[1]	Estimated age, in m.y.[2]
AFRICA	AF1	Damara Belt, Namibia	Moderate	650 (590-720)
	AF2	Shushong Group, Botswana	Small	1875(1750-2000)
	AF3	Ijil Group, Mauritania	Moderate	2100(1700-2500)
	AF4	Transvaal-Griquatown, S. Africa	Very large	2263(2095-2643)
	AF5	Witwatersrand, S. Africa	Small	2720(2643-2800)
	AF6	Liberian Shield, Liberia-Sierra Leone	Large	3050(2750-3350)
	AF7	Pongola beds, Swaziland - S. Africa	Moderate	3100(2850-3350)
	AF8	Swaziland Supergroup, Swaziland S. Africa	Small	3200(3000-3400)
AUS-TRALIA	AU1	Nabberu Basin	Large	2150(1700-2600)
	AU2	Middleback Range	Moderate	2200(1780-2600)
	AU3	Hamersley Range	Very large	2500(2350-2650)
EURASIA	EU1	Altai region, Kazakhstan-W. Siberia	Moderate	375 (350-400)
	EU2	Maly Khinghan - Uda, Far East USSR	Large	550(500-800(?))
	EU3	Central Finland	Moderate	2085 ± 45
	EU4	Krivoy Rog - KMA, USSR	Very large	2250(1900-2600)
	EU5	Bihar-Orissa, India	Large	3025(2900-3150)
	EU6	Belozyorsky-Konski, Ukraine USSR	Moderate	3250(3100-3400)
NORTH AMERICA	NA1	Rapitan Group, NWT Canada	Moderate	700 (550-850)
	NA2	Yavapai Series, Southwest USA	Small	1795(1775-1820)
	NA3	Lake Superior, USA	Large	1975(1850-2100)
	NA4	Labrador Trough and extensions, Canada	Very large	2175(1850-2500)
	NA5	Michipicoten, Canada-Vermilion, USA	Moderate	2725(2700-2750)
	NA6	Beartooth Mountains, Montana, USA	Small	2920(2700-3140)
	NA7	Isua, Greenland	Small	$>3760\pm70$
SOUTH AFRICA	SA1	Morro du Urucum-Mutun, Brazil-Boliva	Moderate	600(?) (450-900)
	SA2	Minas Gerais, Brazil	Very large	2350(2000-2700)
	SA3	Imataca Complex, Venezuela	Large	3400(3100-3700)

[1]See Figure 1

[2]In absence of other data, the assigned age is arithmetic mean of age limits given in brackets.

deformation and erosion. The economically important deposits of El Pao and Cerro Bolivar, Venezuela, represent such remnants; information from these and other scattered localities in the Imataca Complex permit us to infer the approximate original extent of the basin of deposition (nearly 100,000 km^2) and the amount of iron formation deposited (about 10^{13} tonnes).

DISTRIBUTION IN TIME (H.L. James)

The time span encompassed by iron formation deposition is very great - from early Archean to at least as recent as Devonian - and it is unlikely that any geologic era with a duration of as much as a hundred million years will prove to be utterly devoid of such deposits. One persistent association throughout geologic history is that of iron formation with volcanogenic strata of greenstone belts, the position and timing of which

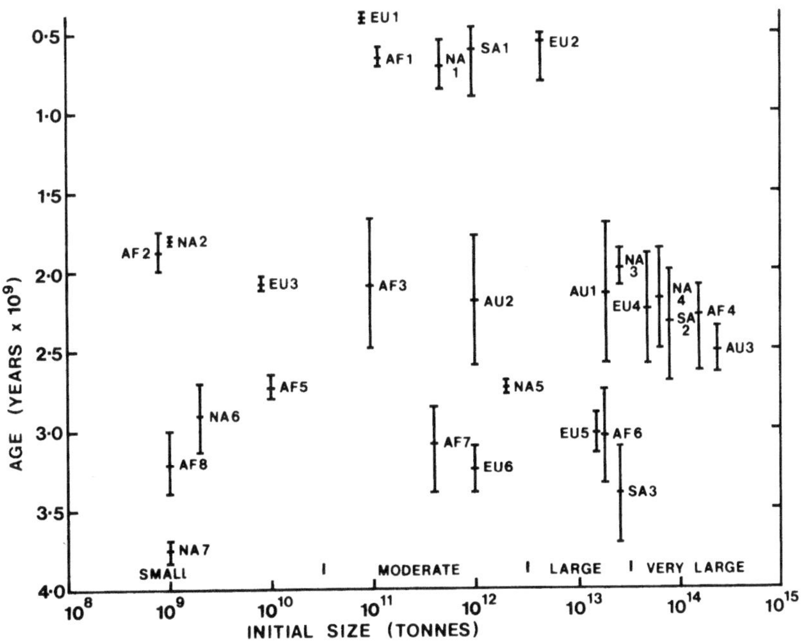

FIG. 1 - Estimated age and size of selected deposits of banded iron formation, from Table 1. Vertical bar indicates possible range in age assignment. Note that tonnage scale is logarithmic. Normal size classes relate to Table 1.

are related to the structural rather than to the chemical evolution of the crust. These deposits, the Algoma class of Gross (16), are particularly common in Archean terranes, but they recur throughout the entire geologic record. These deposits aside, there remains a strong suggestion that the major iron formations, as displayed in Figure 1, tend to be clustered in three age brackets: 500-750 m.y.; 2000-2500 m.y.; and 2900-3400 m.y. The most notable gap in iron formation ages is that between about 750 m.y.a. and 2000 m.y.a. No obvious reason for the general lack of iron sedimentation during this 1250 m.y. period is at hand.

500-750 m.y. Few if any of the deposits assigned to this age bracket in Table 1 are well dated, but the range of uncertainty is not great - none is likely to be younger than 400 m.y. or older than 800 m.y. Two of the listed deposits, EU1 and EU2, appear to be primarily of the volcanogenic association; their location in this time bracket probably reflects simply stages in the orogenic evolution of these particular areas. The remaining deposits, however, together with a number of unlisted noncherty ironstones of other areas, are of particular interest in that each is stratigraphically associated with deposits of possible or probable glacial origin. Young (37) has proposed a direct connection between precipitation of iron and silica and the physicochemical parameters of a glacial environment. Unquestionably, glaciation was widespread in late Proterozoic time on several continents, and although precise age limits are yet to be defined, the association appears to be significant.

2000-2500 m.y. This age bracket doubtless represents the period of greatest accumulation of sedimentary iron in earth's history. All of the districts classed as "Very large" in Table 1, and which in the aggregate contain more than 90% of all known iron-formation, fall within this age bracket. Whether or not deposition occurred within a more limited time span is still uncertain; precise dating of the iron formations, as for Precambrian metasedimentary rocks in general, has proved elusive. Direct measurement using the Pb-Pb method appears to have been applied successfully to the Finnish deposits (27) and possibly to the much older Isua deposits (26), but this approach has not been widely adopted. Most age assignments still rely on the limits set by the age of underlying crystalline rocks and of post-depositional metamorphism or intrusion. The status of age determination for the five principal districts is reviewed briefly below (all Rb/Sr results are adjusted to a ^{87}Rb decay constant of 1.42×10^{11} yr^{-1}).

1. Labrador Trough. A Rb/Sr age of about 1850 m.y. obtained on samples of metasedimentary rocks (14) now is generally believed to represent the time of metamorphism rather than that of deposition. The assigned age of 2175 m.y. is simply the mean of the age of the basement crystalline rock, about 2500 m.y. (2), and that of the metamorphism at 1850 m.y.

2. Hamersley Range, Western Australia. The age of the deposits, once thought to be established at about 2000 m.y. (33), is now believed to be considerably greater. Recent U-Pb analyses of zircons separated from contemporaneous airfall tuff have yielded an age of 2490 m.y. (Compston et al., in preparation). Further limits are set by a Rb/Sr age of 2350 m.y. for later intrusive sills (DeLaeter and Trendall, in preparation), and a 2650 m.y. model age for galena in underlying basalt (Richards et al., in preparation).

3. Quadrilatero-Ferrifero, Minas Gerais, Brazil. No precise age is as yet available for the iron-bearing Minas Group, but on regional grounds it is now considered to be bracketed by a 2700 m.y. age for basement crystalline rocks and an approximately 2000 m.y. age for the Trans-Amazonian orogeny (1).

4. Transvaal-Griquatown basins, South Africa. Shale from the Transvaal Supergroup has been dated by whole rock Rb/Sr analysis at 2263 ± 85 m.y. (18), which may be the approximate age

of deposition of the stratigraphically associated iron formation. A maximum possible age appears to be set by a U-Pb age of 2643 ± 80 m.y. of zircon in underlying lavas (36), and a minimum by a Rb/Sr age of 2095 ± 24 m.y. for a post-Transvaal intrusive (18).

5. Krivoy Rog-KMA, USSR. The age of the iron formations in these adjacent districts is defined essentially by the age of underlying basement rocks (about 2600 m.y.) and the age of post-iron formation metamorphism, which is in the range 1800-2000 m.y. (29). Tougarinov, in an earlier paper dealing with the Aldan Shield ((30), p. 653), mentions in passing a Pb-Pb determination of 2600 m.y. for Krivoy Rog, but no further note of this study has been located. In the absence of more specific information, the assigned age - 2250 m.y. - is the mean of the limits.

2900-3400 m.y. Several major districts fall within this age category, notably those of Bihar and Orissa, India, the Liberian Shield of western Africa, and the Imataca Complex of Venezuela. Age limits for the deposits in the latter two areas (which presumably were continuous prior to opening of the Atlantic in Mesozoic time) are very wide (17,19), and it is not certain whether one or several ages of deposition are represented. The possible age range of the Iron Ore Supergroup of Bihar and Orissa, on the other hand, is both narrower and more firmly established: definite limits are set by Rb/Sr ages of about 3150 m.y. for the "Older Metamorphic Group" and about 2900 m.y. for the later Singbhum Granite (28).

COMPOSITIONAL VARIATIONS (H.L. James)

Quantitative chemical data of reliable quality and significance are not adequate to assess fully either the extent of chemical variations between iron formations of the same general age or of changes in the composition of iron formation as a function of time of deposition. The major deposits of mid-Precambrian-age for which data are available, however, show only a small range in the concentration of the major constituents. Composite chemical analyses for the Mesabi Range, the Labrador Trough, the Hamersley Basin, the Transvaal Basin, and Minas Gerais yield the following averages (in weight percent, range shown in parentheses):

SiO_2	-	45.21 (40.71-48.50)
Fe	-	30.00 (24.46-37.90)
Mn	-	0.22 (0.08-0.49)

It should be noted that in detail the manganese content is much more variable than suggested by these figures: in some districts, such as the Cuyuna Range of Minnesota, the iron formation as a whole may contain 5% or more Mn, and elsewhere, as in the Transvaal Basin, the formation may contain mineable beds of manganese ore. Significant variations may also exist in the amount of sodium; in most districts the Na_2O content, commonly reflected by the presence of the Na-rich amphibole riebeckite in the iron formation, is less than 0.1%, but in the Hamersley Basin it is 0.27% and in the Transvaal Basin it is 1.05%.

There is some suggestion that iron formations of Archean age have a greater range in compositions than those of the mid-Precambrian. Some appear notably higher in silica content and in fact may grade into rock more properly classed as ferruginous chert or jasper. Nevertheless, where systematic data are available, compositions do not vary appreciably from those given for the major mid-Precambrian deposits, as indicated by the data for two typical districts:

	Atlantic City, Wyoming	Moose Mountain, Ontario
SiO_2	44.13	45.20
Fe	33.15	36.70
Mn	0.07	0.04

The younger iron formations, notably those of late Precambrian age associated with glaciogenic deposits, differ substantially in both character and bulk composition from the iron formations of older age. Some are chert-banded, others consist of, or grade into, highly ferruginous, non-cherty maroon shales. The estimated content of major constituents of two of the largest deposits of banded cherty iron formation of this age are as follows, in weight percent:

	Rapitan, Canada	Morro du Urucum, Brazil
SiO_2	25	23
Fe	46	51
Mn	0.08	0.1

These two deposits clearly are more iron-rich than the older formations, but whether this holds true generally for deposits of late Precambrian age is not known.

Few generalizations can be made concerning compositional changes with time, and none can be made firmly. The oldest deposits, those of Archean age, may tend to be more silica-rich than the great mid-Precambrian formations, whereas the late Precambrian deposits may be lower in silica and higher in iron. Manganese content appears to bear little relation to time of deposition; workable manganese ore deposits occur in or closely associated with iron formations of all ages.

PALEOENVIRONMENTAL SIGNIFICANCE (A.F. Trendall)

Attention here is restricted to the cherty banded iron formation: the oxide facies of James (20). This rock-type is distinctive, readily recognized in the field, and effectively defined by a total Fe content between 24 and 35% and a silica content near 45%. Iron oxides and silica together constitute over 90% by weight of the rock. The main mineral components - magnetite, hematite, and finely crystalline quartz (usually called chert) - are so disposed as to define iron-rich and iron-poor layers, beds, or bands of varying thickness. Cherty banded iron formation forms by far the most abundant and significant part of all the rocks termed iron formation. Cherty banded iron formation is divisible into two distinct types, which for lack of agreed and precise nomenclature are referred to here as banded iron formation (abbreviated to BIF) and granule iron formation. Brief descriptions of both types precede discussion of paleoenvironment.

BIF is characteristically banded on several scales. The most detailed description of such banding is that of Trendall and Blockley (35) for the Dales Gorge Member of the Hamersley Basin.

In that rock, regularly spaced millimeter-scale concentrations of iron-bearing minerals define microbands within centimeter-scale mesobands of chert, which are separated by mesobands of iron-rich material referred to as chert-matrix. Higher orders of stratigraphic regularity are also present and are summarized in Figure 2. All of these, including the microbanding, are likely to be, or to have been, laterally continuous over the entire estimated 150,000 km^2 original depositional area of the basin. The general similarity of many features of more strongly altered, or less well exposed, BIF bodies to those of the Hamersley Basin strongly suggests that these presently serve as the best available example on which to base a model for BIF paleo-environmental reconstruction.

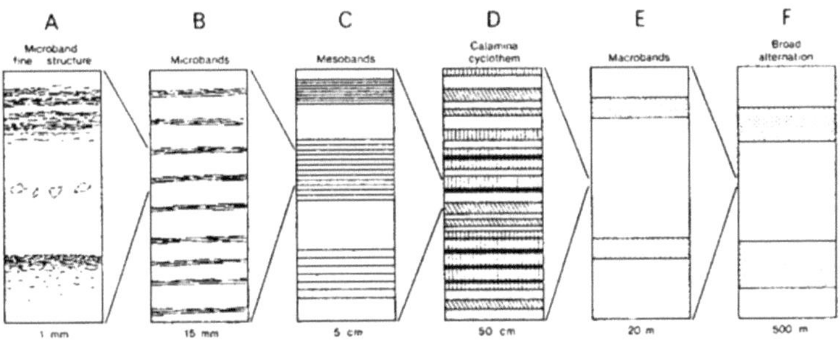

FIG. 2 - Summary of stratification scales within BIFs of the Hamersley Basin. A shows fine structure within the microbands, defined by iron-rich minerals, shown in B. In C, microbanded chert mesobands of different microband interval are represented, separated by chert-matrix (blank). D shows a regular cyclicity of different chert types which occurs within the larger cyclicities shown in E and F.

Although the granule iron formations typically have chemical and mineralogical compositions close to those of BIF, they have distinctively different textures, which have been best described from the Sokoman Iron Formation of Labrador by Dimroth and Chauvel (10) and Zajac (38). In addition to the "granules" - ferruginous ellipsoids about 0.5 mm across within a chert matrix - these rocks display a broad range of elastic sedimentary textures identical with those of many ancient and modern limestones and sandstones. Nevertheless, a cherty banding - the "wavy bedding" of some of the Lake Superior ranges - is often present.

The best geological evidence for the paleoenvironmental significance of cherty banded iron formations is likely to come first from the best preserved and closely studied basins and second from a comparison of the common characters of all known deposits. A theoretical chemical approach to this paleoenvironmental problem also has much of value to contribute but is not within the purview of this paper. This discussion of paleoenvironment begins with a summary of the depositional model for BIF which is most consistent with the BIF of the Hamersley Basin; it is condensed from that presented by Trendall (34) and is largely based on the nature of the banding (Figure 2) and its vertical and lateral regularity. Microbands are regarded as the only primary depositional features within the BIF, and the regularity of microband spacing within any one chert mesoband is taken to indicate that the defining iron-rich/iron-poor cyclicity is related to an equally regular depositional control. Only two natural events involving repeated and widely contrasted environmental changes have great regularity: the day and the year. The general order of microband thickness precludes their diurnal control and it is concluded that microbands are probably seasonally controlled, annual layers: they are varves, and any depositional model must allow for a close relationship between deposition and regular seasonal climatic contrasts. The origin of the fine structure within microbands is not understood. The regional continuity of microbands precludes any depositional mechanism other than chemical precipitation. It is inconceivable that an undisturbed layer of an even thickness no greater than a few millimeters could be spread over an area of at least 50,000 km^2 by a known physical means, unless by the infall of dust, which is considered improbable. The sheer bulk of BIF laid down in the basin makes it unlikely that gross chemical modification has taken place, and it is therefore accepted that both silica and iron were almost the only stable constituents of the deposited material. If this is so, then its origin as a precipitate from the basin water seems more acceptable than an origin as either volcanic or terrigenous dust (6).

The lack of mechanical sedimentary structure in the BIFs indicates an exceptionally still and quiet subaqueous environment. This is one of the most clearly established features of the paleoenvironment and may be related to one (or more) of three reasons: very deep water, a physical surface buffer to weather, or exceptional atmospheric calm. Although ice, or an algal raft, have been considered as possible surface buffers, there is no positive supporting evidence for the existence of either. Trendall and Blockley (35) argue for a water depth of 150-250 m and accept the possibility of an algal raft. There is no evidence for the supply of any terrigenous debris during deposition of the Hamersley Basin BIFs; in uniformitarian terms, the simplest explanation of this seems to be that the climate over the land areas around the basin was completely arid, so that no rivers bearing clastic debris flowed into it. A desert climate, with deposition linked to evaporation, would be consistent also with intrabasinal thickness variations of the BIFs, which thin slightly but consistently away from the center of the ovoid basin, and with the requirement for a close link between deposition and seasonal contrasts. In one BIF of the Hamersley Basin (Weeli Wolli Formation), there is a systematic 23.3-microband cyclicity which is well explained as a reflection of seasonal climatic control.

Trendall and Blockley (35) presented a hypothesis for the diagenetic development of mesobanding by intensification of slight secular variations in primary composition during the dehydration and diagenesis of a compacting sequence of gelatinous, water-rich, colloidal, annual, iron-silica layers. This hypothesis has no immediate paleoenvironmental relevance, except that, if true, it precludes the direct use of mesobanding as evidence for primary depositional environment. Thus the essential paleoenvironmental requirements from geological evidence are the existence of a stable depositional basin of exceptional tranquillity in which climatically controlled chemical precipitation took place in sensitive response to seasonal influences. This stability persisted for at least some millions of years; the absence of stratigraphic signs of desiccation, the need to maintain a supply of iron and silica in solution without the associated clastic debris to be expected in rivers, all argue for some open ocean connection for the basin.

No geological evidence from Hamersley Basin BIFs provides constraints on models for chemical mechanisms of deposition. Ewers (13) has discussed these from a theoretical viewpoint and has concluded that divalent iron in solution in the basin water was the most likely immediate source for the precipitated iron. Ewers suggests that precipitation was triggered either by inorganic photochemical oxidation or by oxygen made available by green plant photosynthesis; an essentially anoxic atmosphere is required. The possible presence of microfossils (22,25) and of isotopically light carbon in carbonates (3) supports the presence of plant life in the basin, but the evidence is by no means conclusive.

Turning now to BIF deposits other than those of the Hamersley Basin we find that the principal features which there form the basis of paleoenvironmental interpretation - the presence of iron-rich and iron-poor mesobands, sometimes microbanded, with a high degree of lateral continuity - are universal features of other BIFs; by implication, an environment confidently established for the deposits of the Hamersley Basin could not be discounted elsewhere. An exception to be noted is the association of BIFs with glacigene sediments, which appears contrary to Trendall's suggestion (32) of an evaporitic initiation of precipitation. Young (37) has reviewed this association and described one fine example in the Rapitan Group of Canada. However, as Young pointed out, the precise paleoenvironmental significance of the association is speculative.

The source of the parent materials (and especially the iron) for iron formation deposition has been a long-standing problem, which has indirect relevance for paleoenvironmental interpretation. The two main alternative hypotheses, those involving a close volcanic source and those which involve derivation of iron by weathering, have each been ably argued; Eichler (12) gives a good review. We accept that, except for the smaller deposits clearly relatable to associated volcanics, there is no apparent need to invoke volcanic activity directly in any paleoenvironmental interpretation of iron formation. As far as the weathering hypothesis is concerned, it has been shown that its application generates more problems than it solves (31), particularly if chemical models for iron concentration of the simplicity of that of Drever (11) are considered adequate.

The depositional environment of BIF has been discussed before that of granule iron formation because the paleoenvironmental significance of the latter is a more complex problem. It seems generally agreed that granule iron formation represents a high-energy, essentially clastically reworked, deposit. To this extent its paleoenvironmental significance is uncontroversial. However, there are widely opposed published viewpoints as to the origin and nature of the material which was reworked.

For the Sokoman Iron Formation, Zajac ((38), pp. 33-36) notes a range of possibilities for the origin of granules (and the larger associated oolites), including simple rounding of fragments of penecontemporaneously precipitated ferruginous chert, direct precipitation as gelatinous globules, and some form of organic growth. Dimroth and Chauvel ((10), p. 117) accept a similar choice of hypotheses; although they emphasize the textural affinities of the Sokoman Iron Formation with many limestones, they specify that "iron formations are <u>not</u> limestones that were replaced by chert and iron compounds after deposition"; that is, the primarily precipitated material <u>was</u> silica- and iron-rich. However, Dimroth (8,9) has subsequently leaned towards the hypothesis of Kimberley (23) that (?all) iron formations result from the diagenetic replacement of aragonite mud. Apart from the negative objection that the main evidence for this hypothesis is the textural similarity of limestone and granule iron formation, we find it impossible to believe that the very large deposits of iron formation could have been formed by a post-depositional chemical transformation that shows no sign of gradational stages, that is nowhere stratigraphically discordant, and that leaves no surviving relics of unreplaced parent material. This question must be resolved before the problem of the paleoenvironmental significance of granule iron formation can be addressed.

REFERENCES

(1) Almeida, F.F.M. 1978. Chronotectonic boundaries for Precambrian time division in South America. Acad. Brasil Cienc. Ar. 50: 527-535.

(2) Baadsgaard, H.; Collerson, K.D.; and Bridgwater, D. 1979. The Archean gneiss complex of northern Labrador. I. Preliminary U-Th-Pb geochronology. Can. J. Earth Sci. 16: 951-961.

(3) Becker, R.H., and Clayton, R.N. 1972. Carbon isotopic evidence for the origin of a banded iron-formation in Western Australia. Geochim. Cosmochim. Acta 36: 577-595.

(4) Brandt, R.T.; Gross, G.A.; Gruss, H.; Semenenko, N.P.; and Dorr, J.V.N. 1972. Problems of nomenclature for banded ferruginous-cherty sedimentary rocks and their metamorphic equivalents. Econ. Geol. 67: 682-684.

(5) Bronner, G., and Chauvel, J.J. 1979. Precambrian banded iron-formations of the Ijil Group (Kediat Ijil, Requibat Shield, Mauritania). Econ. Geol. 74: 77-94.

(6) Carey, S.W. 1976. The Expanding Earth. Developments in Geotectonics, vol. 11. Amsterdam: Elsevier.

(7) Coertze, F.J.; Burger, A.J.; Walraven, R.; Marlow, A.G.; and MacCaskie, D.R. 1978. Field relations and age determinations in the Bushveld Complex. Geol. Soc. South Africa Trans. 81: 1-11.

(8) Dimroth, E. 1976. Aspects of the sedimentary petrology of cherty iron-formation. In Handbook of Stratabound Ore Deposits, ed. K.H. Wolf, vol. 7, pp. 83-88. Amsterdam: Elsevier.

(9) Dimroth, E. 1977. Facies Models 6. Diagenetic facies of iron formation. Geosci. Can. 4: 83-88.

(10) Dimroth, E., and Chauvel, J.J. 1973. Petrography of the Sokoman Iron Formation in part of the central Labrador Trough, Quebec, Canada. Geol. Soc. Am. Bull. 84: 111-134.

(11) Drever, J.I. 1974. Geochemical model for the origin of Precambrian banded iron formations. Geol. Soc. Am. Bull. 85: 1099-1106.

(12) Eichler, J. 1976. Origin of the Precambrian banded iron-formations. In Handbook of Stratabound Ore Deposits, ed. K.H. Wolf, vol. 7, pp. 157-201. Amsterdam: Elsevier.

(13) Ewers, W.E. 1980. Chemical conditions for the precipitation of banded iron-formations. 4th International Symposium on Environmental Biogeochemistry, Canberra, Australia Academy of Science.

(14) Fryer, B.J. 1972. Age determination in the Circum-Ungava Geosyncline and the evolution of Precambrian banded iron-formations. Can. J. Earth Sci. 9: 652-663.

(15) Goldich, S.S. 1973. Ages of Precambrian banded iron-formations. Econ. Geol. 68: 1126-1134.

(16) Gross, G.A. 1965. Geology of iron deposits in Canada, vol. 1. General geology and evaluation of iron deposits. Canada Geol. Survey Econ. Geol. Rept. 22, p. 181.

(17) Gruss, H. 1973. Itabirite iron ores of the Liberia and Guyana shields. In Genesis of Precambrian Iron and Manganese Deposits. UNESCO Earth Science Series, vol. 9, pp. 335-359.

(18) Hamilton, P.J. 1977. Isotope and trace element studies of the Great Dyke and Bushveld mafic phase and their relation to early Proterozoic magma genesis in southern Africa. J. Petrology 18: 24-53.

(19) Hurley, P.M.; Fairbairn, H.W.; and Guadette, H.E. 1976. Progress report on Early Archean rocks of Liberia, Sierra Leone, and Guyana, and their general stratigraphic setting. In The Early History of the Earth, ed. B.F. Windley, pp. 511-521. New York: John Wiley and Sons.

(20) James, H.L. 1954. Sedimentary facies of iron-formations. Econ. Geol. 49: 235-293.

(21) James, H.L. 1966. Chemistry of the iron-rich sedimentary rocks. U.S. Geol. Survey Prof. Paper 440-W.

(22) Karkhanis, S.J. 1976. Fossil iron bacteria may be preserved in Precambrian ferroan carbonate. Nature 261: 406-407.

(23) Kimberley, M.M. 1974. Origin of iron ore by diagenetic replacement of calcareous oolite. Nature 250: 319-320.

(24) Kimberley, M.M. 1978. Paleoenvironmental classification of iron formations. Econ. Geology 73: 215-229.

(25) La Berge, G.L. 1967. Microfossils and Precambrian iron-formations. Geol. Soc. Am. Bull. 78: 331-342.

(26) Moorbath, S.; O'Nions, R.K.; and Pankhurst, R.J. 1973. Early Archaean age for the Isua iron formation, West Greenland. Nature 245: 138-139.

(27) Saako, M., and Laajoki, K. 1975. Whole rock Pb-Pb isochron age for the Pääkkö iron formation in Väyrylänkylä, South Puolanko area, Finland. Geol. Soc. Finland Bull. 47: 113-116.

(28) Sarkar, S.N.; Saha, A.H.; Boelrijk, N.A.I.M.; and Hebeda, E.H. 1979. New data on the geochronology of the Older Metamorphic Group and the Singhbhum granite of Singhbhum-Keonjar-Mayurbhanj region, eastern India. India J. Earth Sci. 6: 32-51.

(29) Semenenko, N.P. 1973. The iron-chert formations of the Ukranian Shield. In Genesis of Precambrian Iron and Manganese Deposits. UNESCO Earth Science Series, vol. 9, pp. 135-142.

(30) Tougarinov, A.I. 1968. Geochronology of the Aldar Shield, southeastern Siberia. Can. J. Earth Sci. 5: 649-656.

(31) Trendall, A.F. 1965. Origin of Precambrian banded iron formations (Discussion). Econ. Geol. 60: 1065-1070.

(32) Trendall, A.F. 1973. Iron-formations of the Hamersley Group of Western Australia: type examples of varved Precambrian evaporites. In Genesis of Precambrian Iron and Manganese Deposits, Proceedings of the Kiev Symposium, Unesco, Paris. Earthsciences 9: 257-269.

(33) Trendall, A.F. 1973. Precambrian iron-formations of Australia. Econ. Geol. 68: 1023-1034.

(34) Trendall, A.F. 1976. Geology of the Hamersley Basin. 25th International Geology Congress, Sydney, Australia, Excursion Guide 43A.

(35) Trendall, A.F., and Blockley, J.G. 1970. The iron formations of the Precambrian Hamersley Group, Western Australia. Geol. Survey West. Australia Bull. 119: 1-366.

(36) Van Niekerk, C.B., and Burger, A.J. 1978. A new age for the Ventersdorp acidic lavas. Geol. Soc. S. Africa, Trans. 81: 155-163.

(37) Young, G.M. 1976. Iron-formation and glaciogenic rocks of the Rapitan Group, Northwest Territories, Canada. Precambrian Res. 3: 137-158.

(38) Zajac, I.S. 1974. The stratigraphy and mineralogy of the Sokoman Iron Formation in the Knob Lake area, Quebec and Newfoundland. Geol. Surv. Can. Bull. 220: 1-59.

Mineral Deposits and the Evolution of the Biosphere, eds. H.D. Holland and M. Schidlowski, pp. 219-236. Dahlem Konferenzen, 1982.
Berlin, Heidelberg, New York: Springer-Verlag.

Variations in the Distribution of Mineral Deposits with Time

R. E. Folinsbee
Dept. of Geology, University of Alberta
Edmonton, Alberta T6G 2E3, Canada

Abstract. Volcanogenic base metal deposits and related precious metal deposits were formed in the crust at least as long as 2700 m.y. ago, at the time of craton formation. The Rhodesian and Kaapvaal cratons were the source of the immense gold and important uranium deposits of the Witwatersrand basin (2700 m.y.); the Superior craton of Canada was the source of uranium in the Blind River uranium deposit (∿2400 m.y.). These placer deposits carry detrital pyrite and appear to have been eroded and transported under an oxygen-deficient atmosphere. The principal period of banded iron formation occurred 2500-2000 m.y. ago at a time when sufficient oxygen had accumulated in the ocean-atmosphere system to bring about iron oxide precipitation in favorable basins. Stratiform lead-zinc deposits in sediment-hosted environments formed in mobile belts marginal to the cratons 2000-1400 m.y. ago. About the same time (1800-1100 m.y.), uranium deposits, sometimes with associated gold, silver, copper, nickel, or cobalt, were forming in the Cahill basin of Northern Australia and in the Athabasca basin of Canada. Sedimentary copper deposits of the Zambian or Kupferschiefer type developed from 1400 m.y. to 200 m.y. ago. The setting of these deposits suggest extensive oxidation and red bed conditions on the neighboring land masses. Economic phosphorites are late Precambrian through Phanerozoic in age and show clear biospheric control. Volcanogenic base metal deposits reappear in the Paleozoic Appalachian orogenic belt at Bathurst, Canada and in the Miocene of the Kuroko district of Japan. Mississippi Valley lead-zinc deposits occur in platform carbonates from late Precambrian (Nanisivik, Gayna River) through Paleozoic (Missouri, Polaris, Pine Point) into Mesozoic times (Silesia). Porphyry coppers of late Phanerozoic age occur for the most part in subduction-related volcanic belts of the Circum-Pacific and Alpide orogenic regions. A recently recognized class of ultra-fine gold-silver deposits occurs in Phanerozoic areas of former hot spring activity, formed by the interaction of meteoric waters with shallow seated igneous intrusives.

INTRODUCTION

The unequal distribution of mineral deposit types in time appears to be due to the interplay of processes related to the nature of the earth's mantle, its heat flow and convective motions, with atmospheric processes, the evolution of the earth's oxygen budget, and life (6). Four and half billion years of earth history have produced enough large, rich and accessible deposits to have met the needs of preindustrial society for eight millenia, and the needs of the industrial world for a little over two centuries. At the beginning of a period of shortages in energy and materials, enough information on the ore deposits of the world must be available so that an intelligent search can be made for those deposits that have escaped the eye of the explorationist.

In this paper on variations in the nature of mineral deposits with time, I propose to deal with the principal groups of deposits in a chronological framework and to demonstrate that even in the development of volcanogenic deposits of deep seated origin, the atmosphere (as an oxidizing agent, acting through the medium of meteoric water) plays some part. Oxygen in the atmosphere-ocean system was critical in the formation and enrichment of the banded iron formations; since these deposits are covered thoroughly in the paper by James and Trendall (this volume), I shall only touch on them briefly.

A role for groundwater in the formation of uranium deposits is increasingly recognized. Uranium deposits (with associated gold, nickel, and copper) are clearly related to the Zambian or Kupferschiefer type copper (with associated nickel, cobalt, uranium) deposits. Ground waters play an important role in the karsting which precedes emplacement of Mississippi valley type lead-zinc deposits, and cold sulfur springs of meteoric origin provide temperature gradients and ingredients necessary for precipitation. The part played by ground water in hot spring deposits is well-known; the importance of the environment as a host to fine-grained gold and silver deposits of the Carlin type has only recently been recognized. Great mining companies such as Newmont and Homestake are now in the forefront of exploration for this type of deposit. Even the 900 centimeter annual rainfall at OK Tedi in the remote regions of Papua-New Guinea is significant in this story. Meteoric waters entered into the volcanic system that produced the alteration zones and porphyry copper-gold protore and brought about the erosion and secondary enrichment that exposed the ore body and raised the gold content of the residual gossan to a profitable level.

I shall not deal with purely magmatic ore deposits such as copper, nickel, platinum, chromium, vanadium, and tin deposits of the Sudbury and Bushveld lopoliths. Nor shall we concern ourselves with the great stratiform lead-zinc-silver-copper ore bodies in sedimentary environments. Most of the stratiform deposits, though formed on the sea floor, seem to have been deep-seated in origin; the hydrothermal fluids probably rise along graben-like megasutures to give us the great ore bodies at Broken Hill and Mt. Isa in Australia, Sullivan in Canada, and the newly discovered lead-silver ore bodies of the Northern Cape district of South Africa. The relationship between these ore bodies and the nature of the contemporary atmosphere and hydrosphere is tenuous at best.

VOLCANOGENIC MASSIVE SULFIDES AND RELATED PRECIOUS METAL DEPOSITS

Hutchinson (14) and Sangster (24) have given us excellent review papers of this subject; atmospheric evolution has comparatively little to do with this type of deposit, and I shall touch on those few aspects of these deposits that are related to the evolution of the atmosphere and biosphere. It is now widely recognized that Kuroko type deposits originate as fumarolic deposits introduced into the sea or into a volcanic caldera, usually at considerable depth. It was only last year that we were able to see on public television, under the auspices of the National Geographic Society, hydrothermal plumes ejected from sulfide chimneys located along the East Pacific Rise at 21°N (8). The accompanying bizarre fauna of clams and giant worms lends an exotic touch to these deep sea deposits. This

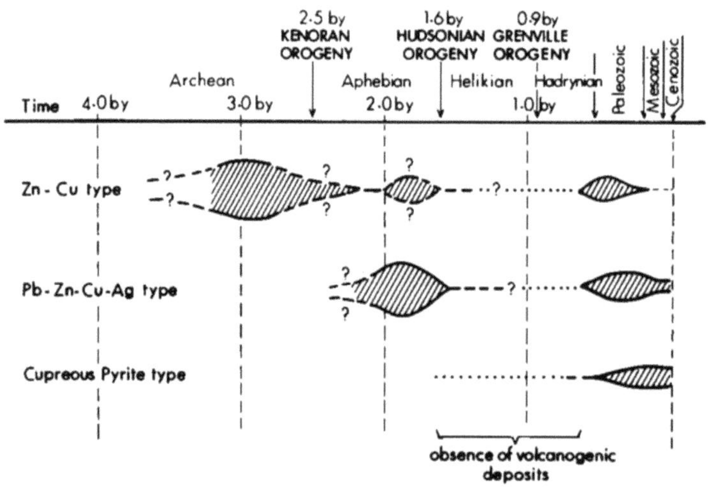

2·5 by
KENORAN
OROGENY
1·6 by
HUDSONIAN
OROGENY
0·9 by
GRENVILLE
OROGENY
Archean
Aphebian
Helikian
Hadrynian
Paleozoic
Mesozoic
Cenozoic
Time
4·0 by
3·0 by
2·0 by
1·0 by
Zn - Cu type
Pb - Zn - Cu - Ag type
Cupreous Pyrite type
absence of volcanogenic deposits

of carbon dioxide have been added to the ocean-atmosphere system by carbon dioxide-rich fluids at this time.

The Timmins camp has been Canada's primary gold producer; it can scarcely be accidental that Kidd Creek with 20,000 tonnes, i.e., 600,000,000 ounces, of silver reserves and Timmins, which has produced 2,000 tonnes of gold, are so closely related in time and space. The ancient Rhodesian and Kaapvaal cratons of Southern Africa are host to important gold mines of the Timmins type, but so far they have not been major base metal producers. The massive sulfide deposits of the Paleozoic Bathurst Camp in Canada and the Miocene Kuroko district in Japan are volcanogenic sea floor deposits associated with basic and felsic volcanic piles related to island arcs; the latter is the type district for these deposits. There is very little difference between Archean, Paleozoic, and Cenozoic massive sulfide deposits of the volcanogenic type. With time there is a little more lead with a larger fraction of the radiogenic lead isotopes, and more sulfate gangue (barite, anhydrite) possibly reflecting more oxidizing conditions at the sea floor depositional site.

MINERAL DEPOSITS WITH ATMOSPHERIC OR HYDROSPHERIC CONTROLS

Archean and Proterozoic Placer Gold and Uraninite Deposits

While placer gold occurs in deposits of all ages, the association of placer gold with uraninite is confined to the Archean and early Proterozoic; the only known exception is a minor placer uraninite occurrence along the Indus drainage. Pyrite accompanying the gold and uraninite of the Witwatersrand basin conglomerates (reefs) is clearly detrital, in places retaining a rounded or buckshot form. Since both pyrite and uraninite perish in an oxidizing environment, it is a reasonable conclusion that the atmosphere of the Archean and earliest Proterozoic was deficient in oxygen.

Goldfields, deriving their gold and uranium from the precious metal deposits of the Kaapvaal and Rhodesian cratons, developed around entry channels on the rim of the Witwatersrand basin (7). Pretorious (20) has given us an admirable view of the depositional environment of the gold-fields and has reviewed speculations and observations of this largest repository of the world's gold from which has come 55% of all the gold that has been recovered by man since the beginning of civilization. The basin filled from the northwest during the time from 3000 m.y. ago to 2600 m.y. ago. Earlier gold and uranium bearing conglomerates (reefs, in South African mining parlance) were exposed and eroded to reconcentrate gold and uranium in the upper reefs. In front of the building housing the Economic Geology unit at the University of Witwatersrand there is a large block of this ore, containing rounded pebbles of placer pyrite, which are now rapidly altering to iron oxide in the Johannesburg smog. The complete absence of evidence of oxidation or of uranium movement in solution in the reefs of the Rand system is persuasive evidence of low levels of oxygen in the atmosphere at the time of deposition.

These low levels of oxygen may have persisted into lower Proterozoic time, for in the Blind River district of Ontario we find pebble conglomerates like those of the Witwatersrand carrying placer pyrite and enough uraninite and brannerite to give large tonnages of uranium ore of modest grate (0.15% U_3O_8). Though gold can be traced throughout the Proterozoic paleochannels that developed on the Superior craton, no gold ore deposits of Rand caliber have yet been located.

In 1975 I was greatly impressed by the filamentous and finger-like structures containing fine gold preserved in the thucolitic carbon reef structures characteristic of the border facies of the Rand deltaic fans. Hallbauer(9) thought these filaments resembled algae or lichen, and that they had acted as a trap for fine sedimentary gold. Though Cloud(3) has thrown a shadow over this intriguing picture by producing similar structures inorganically, Hallbauer (10) has presented additional evidence suggesting that the Witwatersrand Carbon Reef and Carbon Reef Leader originated in a biological environment of algal/bacterial mats, now preserved as bituminous coal-like seams. Zumberge, Sigleo, and Nagy (29) have made a definitive molecular and elemental analysis of this carbonaceous material and concluded that it is (or was) kerogen produced by polymerization of biochemicals from decayed primitive Precambrian microorganisms.

Iron Ores Derived from Banded Iron Formations (BIF's)

The Lake Superior type Proterozoic iron formations and related ores have been the mainstay of 20th century steel making, and the deposits have come in for intensive study since the first reports of Van Hise at the turn of the century. Since banded iron formations are covered authoritatively by James and Trendall (this volume), they need not be considered in detail here. The reader might also wish to refer to the beautifully illustrated review paper on the Transvaal and Hamersley basins by Button (2). Holland (12) holds that the only reasonable source for the immense concentration of iron and phosphorus in these banded iron formations is seawater, perhaps anoxygenic at depths and capable of holding ferrous iron which is brought to the surface in areas of upwelling.

There is much controversy about the cause for upwelling; Young suggests that it may have been glacially induced. A glacial peak is associated with the beautifully banded 800 m.y. Rapitan iron ore of the Snake River area, northwestern Canada. Dropstones occur within the banded iron formation, which contains 45% Fe, an unusually high concentration of the metal. This deposit, though remote from sea and civilization, was considered seriously by California Standard Oil Company as a major source of iron after its discovery and evaluation in the early 1960s. The later, sensational developments in the Hamersley basin have precluded development of the Rapitan deposit (Clint Dahlstrom, California Standard, personal communication).

Phosphorites

Phosphorus bears somewhat the same relationship to agriculture that iron does to industry; both are elements of primary importance, won for the most part from sedimentary deposits of limited areal extent but great richness. The March 1979 issue of Economic

Geology was devoted to phosphate, potash, and sulfur. The paper on reevaluation of the spatial and temporal distribution of phosphate deposits in the light of plate tectonics by Cook and McElhinny (4) was particularly pertinent to this conference. I include only a figure (Fig. 2) from the Cook-McElhinny paper as the topic was a major focus of discussion in Berlin and is considered in another part of this volume.

Proterozoic Uranium Deposits (Associated Ni, Co, Cu, Au, Ag)
During the early part of the 1970s, the price of uranium rose sharply in response to the limited supply of the metal and an expected burgeoning of the demand for uranium for power reactors. This prompted the Rio Tinto group to bring Rössing, a low grade (350 ppm uranium) pegmatite in southwest Africa (Namibia), into production. At the same time, Rio Algom, a daughter company,

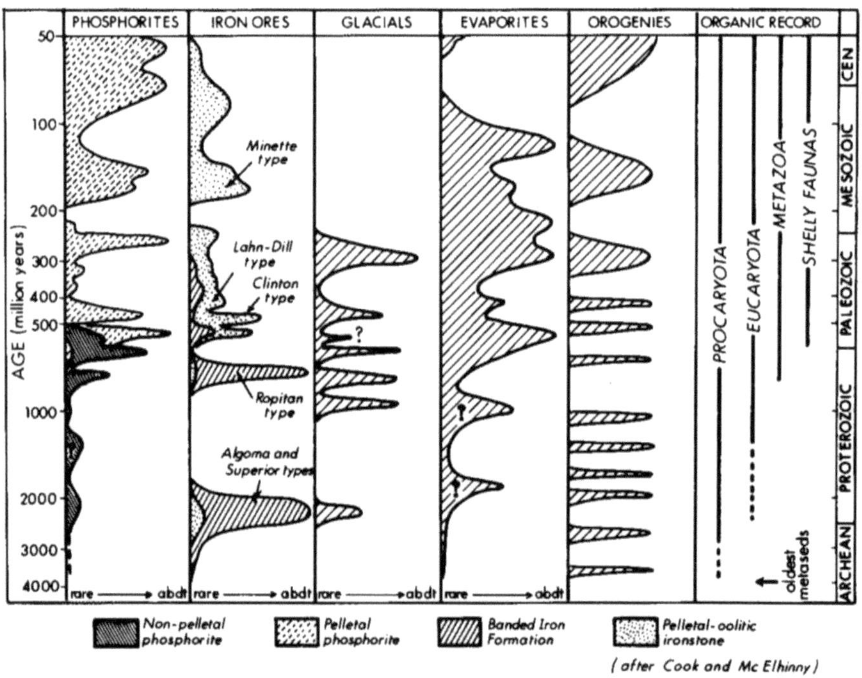

FIG. 2 - The temporal distribution of phosphorites and sedimentary, organic, and tectonic events which may influence phosphogenesis (4).

commenced recommissioning marginal mines in the Blind River camp with attendant high costs. World-wide prospecting for uranium, largely by petroleum companies, was carried forward on an unprecedented scale. Spectacular finds of uranium were made in the Northern Territory of Australia, in the Athabasca basin of Saskatchewan, Canada (7), and most recently at Olympic Dam in South Australia.

The Cahill basin of the Northern Territory of Australia is probably a euxenic, carbon-rich back reef lagoon which may have derived its uranium from the weathered and eroded uranium-rich Nanambu granites of the neighboring craton, though a marine source has also been invoked. Reserves are large (400,000 tonnes of uranium oxide), and the potential resource is great. The Olympic Dam prospect in South Australia may contain even more uranium, for preliminary drilling has outlined more than 500 million tonnes of uranium- and gold-bearing copper ore (11), and the find is far from delineated.

Reserves in the Athabasca Basin of Northern Saskatchewan, Canada are about 200,000 tonnes of uranium oxide and are growing rapidly. Jan Hoeve, the 1980 Distinguished Lecturer of the Society of Economic Geologists, recently presented a paper on uranium metallogenesis in Saskatchewan, reflecting on the role of the basement, unconformity, and Athabasca formation. He believes that the uranium was derived from uranium-rich granites and supracrustal sediments and that uranium, together with nickel and gold, moved for long distances in oxidizing waters of meteoric origin. The solutions are thought to have descended, to have been heated by igneous rocks associated with the 1700 m.y. Churchillian orogeny, or the 1100 m.y. period of diabase dike intrusion, or by the distant effects of a late Paleozoic orogeny in the ancestral Rocky Mountains about 300 m.y. ago. Heated, mineralized solutions rose along faults to and through the unconformity at the base of the Athabasca sandstone, where deposition of uranium, gold, silver, and nickel took place in the presence of carbonaceous material.

This combination of elements, with the exception of gold, is reminiscent of the well-known vein-type Ag-Co deposits of Cobalt, Ontario and the U-Ag deposits of Great Bear Lake, N.W.T. Individual ore bodies contain up to 50,000 tonnes of uranium oxide, some with equal amounts of nickel, and in other instances with multigram quantities of gold or silver per tonne. Ore grades average several percent uranium oxide, but pods may contain up to 50% uranium oxide, a spectacular ore. Hoeve draws analogies with the roll front uranium deposits of Wyoming, where uranium is transported in oxidizing solutions down dip in sandstone aquifers by meteoric waters to a front or interface where the solution encounters reducing conditions and deposits a roll front ore body.

Sedimentary Copper Deposits

Rowlands and his colleagues (23) in the conclusion to their paper on Adelaidean stratiform copper deposition hint at a general world-wide pattern of stratiform copper deposition in the period from ca. 1400 m.y. ago to the beginning of the Paleozoic, 600 m.y. ago. Rock sequences of this age are lithologically similar on every continent. Three or more tillites, accompanied by similar shelf carbonates with stromatolitic dolomites lend confidence to world-wide correlations. Stratiform copper may occur in dip-oriented sand facies of deltaic origin as at Udokan, east of Lake Baikal in Siberia, one of the world's largest copper deposits or in strike-oriented shale facies characteristic of many deposits of the Zambian copper belt (22). Abundant red beds began to appear in the geologic section about 1800 m.y. ago, and the red bed environment is generally held to be an indication of arid conditions in tectonically inactive regions with internal drainage (15). Under such conditions, streams would be intermittent, acidic in the manner of the present day Rio Tinto of Spain or the mine waters of Butte, Montana, capable of carrying copper in solution to a neighboring euxenic basin. Here pyrite, chalcopyrite, bornite, and chalcocite might have been precipitated, occasionally with cobalt and uranium.

The Kupferschiefer deposits of the Triassic Zechstein basin of Europe are a well-known Mesozoic example of a copper bearing black shale facies, and the new copper deposits in the Polish sector of the basin are very large indeed. Holland (13) has suggested that the enrichment of a number of elements (Ag, Cr, Cu, Mo, Ni, Pb, V, and Zn) in the black shale facies is proportional to the concentration of these trace metals in seawater, and that this concentration has not changed significantly during Phanerozoic time; the near-constancy may extend throughout the latter part of the Proterozoic. We should, perhaps, look to trace elements in seawater as the major source for the metals in the euxenic black shale facies rather than to a source in neighboring land masses.

North (16,17) and Demaison and Moore (5), in examining the very unequal distribution of prolific oil basins in space and time, hold that this distribution is not so much a function of a capricious preservation of oil as of the episodic nature of the formation of effective source sediments. This may be true for ore deposits as well. Perhaps sources such as uranium-rich granites, copper- or gold-rich volcanic rocks, muds rich in lead and zinc from which a metalliferous brine may be expelled are prerequisites for the development of ore bodies. A second prerequisite is a carrier; this may be seawater, a connate brine, or hydrothermal solutions (which often turn out to be meteoric in origin). The third prerequisite is a trap analogous to an oil reservoir. These may be semipermeable barriers, karsted limestones or dolomites, brine pools in a rifted zone, or sedimentary basins. My feeling that the copper in the mine series of the Zambian copper belt came from neighboring rocks was strengthened in 1974 by seeing cores with 100 foot lengths of primary, chalcopyrite-bearing Nchanga red granite unconformably below the mine series ore bodies. Here was a clear indication that erosion of granite had produced the arkosic quartzites of the mine series; in this case there appears to be a close association of source and trap.

Mississippi Valley Lead-Zinc Deposits

Ohle (18) has recently published a very thoughtful analysis of the widely held basinal brine theory for the deposition of

Mississippi Valley lead-zinc deposits; he ended this paper with a plea that all geologists remain inquisitive and critical. In his paper he discusses the problems of the source beds, timing, character of the solutions, temperatures, and paths of migration, and holds that many of these problems are unresolved. In its simplest form the genetic model of Beales and Jackson (1) suggests that Pine Point, located on the margin of a deep sedimentary basin containing hydrocarbons and evaporites, developed as a lead-zinc deposit because basin waters were expelled during compaction and migrated up-dip through porous Devonian reef until they encountered reduced sulfur in cold waters in a karst system produced by the attack of carbon dioxide-rich waters of meteoric origin. One hundred million tons of 10% Pb-Zn ore were deposited in collapse structures of the cavern system. Cold sulfur-bearing springs still abound in the area. Ohle points out that the Beales-Jackson model fits well at Pine Point, but that problems arise when the model is applied to other districts. Lead at Pine Point is conformable; this suggests deposition from a uniform source about 300 m.y. ago at about the time of karst preparation.

However, most deposits in the Mississippi Valley contain "J"-type radiogenic lead that must have come from a source and by a route quite different from that at Pine Point. Mississippi Valley-type lead-zinc deposits are not confined to the Mississippi Valley nor to the Paleozoic era. At Gayna River in the Northwest Territories of Canada, deposits of this type have turned up in a Proterozoic stromatolitic reef; on Baffin Island they are being mined from a Proterozoic karst cavern in carbonate at Nanisivik; in Silesia they occur in carbonates marginal to the Triassic Zechstein basin.

Porphyry Copper Deposits

The plate tectonic model for the origin of porphyry copper deposits advanced by Sillitoe (25) is widely accepted. In this paper we are not so much concerned with the source of the metal (subducted metal-rich ocean floor sediments and volcanics) or the carrier (a subducting plate and later an ascending body of calc-alkaline magma), as with the trap. The trap was a volcano that interacted with meteoric water to undergo extensive alteration and the formation of a porphyry copper ore body at the core of the alteration zone. Sheppard, Nielsen, and Taylor (26) have effectively demonstrated that meteoric water from local rainfall plays a crucial part in the alteration and emplacement of porphyry copper ore bodies. We are also concerned with erosion, largely by rainwater, required to expose the porphyry copper, and the production of oxide and secondary sulfide zones that may be critical in the economics of mining. I choose as my example the OK Tedi porphyry gold prospect in the remote regions of Papua, New Guinea, 800 kilometers from the sea on the upper reaches of the Fly River at an elevation of 2890 meters and with a rainfall of 900 centimeters.

The progenitor volcano (Mt. Fubilan) was emplaced 2.6 m.y. ago; the subsequent copper mineralization formed 1.1 m.y. ago in association with the Fubilan porphyries (19). It has taken only a million years for the 250 million tonne ore body to be exposed and enriched. Copper has been leached from the gossan (iron cap) of the deposit. Its concentration varies from the 0.85% Cu level of the primary ore to trace amounts; the gold content, however, was increased from 1 to 5 grams per tonne during weathering. Primary ore is worth about $40 a tonne; the gold-rich cap ore is worth $100 a tonne when the price of gold is $600 an ounce. The gold bearing gossan will be the first mining target; in five years it will yield 75 tonnes of gold which are currently worth 1.4 billion dollars, enough to justify the capital expenditure of one billion dollars required for the project (Mining Annual Review, 1980, London, page 432).

Copper porphyry bodies are not confined to the Cenozoic. Important deposits occur in the Triassic Guichon Creek batholith of the Highland valley region of British Columbia, probably preserved by reason of the protection from erosion offered by an extensive columnar basalt cap rock. Gaspé in Québec has a Paleozoic porphyry copper ore body, and a 2700 m.y. old porphyry copper gold ore body occurs deeply buried in the Pearl Lake porphyry at Timmins, Ontario.

Carlin Type Gold Deposits

Radke, Rye, and Dickson (21) have demonstrated conclusively that the Carlin gold deposit of Nevada developed as a result of hydrothermal processes initiated by heat from an underlying Tertiary intrusive, and that the ore forming fluids were entirely of meteoric origin. These fluids received their heat from an igneous mass, reacted with source rocks at depth, extracted ore and gangue minerals, and carried these upwards along steep faults into near-surface, permeable units where the ore was deposited. Immense amounts of meteoric water were involved; some 10 billion tonnes of metoric water were heated to temperatures of 200 to 300°C to form the 10 million tonnes of gold ore at Carlin.

Carlin-type ore bodies in the Piedmont region of the Carolinas and Georgia are associated with shallow rhyolitic intrusives of early Paleozoic age; the deposits were important gold producers in the early days of Appalachian gold mining (28). Since the gold in Carlin-type deposits is typically very fine-grained, it does not pan easily, and the early wave of prospecting throughout the world may have missed many deposits of this class. Homestake Mining this year made a sensational discovery of Carlin-type gold in Carolina.

SUMMARY AND CONCLUSION

It appears that there has been progressive evolution of mineral deposits with time and that the changing nature of the atmosphere, biosphere, and hydrosphere has played some role in this evolution. Change with time is least apparent in the case of the volcanogenic ore deposits of Kuroko type. Evolution is most apparent in the case of sedimentary iron ores and phosphorites; for the latter changes in the biosphere have apparently been critical. Uranium deposits of the Archean and earliest Proterozoic are for the most part fossil placers. Later in the Proterozoic, uranium (an element susceptible to oxidation) became more mobile; along with a rather strange group of partners: gold, silver, nickel, cobalt, and copper, uranium began to move more readily in solutions in the lithosphere and to be deposited near unconformities and in basins. Sedimentary copper deposits appear in abundance in the later part of the Proterozoic, sometimes in delta sand facies, as at Udokan, or in the shale facies, as in Zambia and Zaire. Associated uranium, cobalt, and nickel in these copper deposits suggest a close relationship with the sedimentary uranium deposits. Mississippi Valley-type lead-zinc deposits had to await the advent of the extensive platform carbonates of the later Proterozoic and Phanerozoic Eras. Meteoric waters as well as connate brines played a role in their formation. Porphyry coppers are related to subduction zones; they belong largely to the era of plate tectonics; some porphyry copper deposits are very young; their formation must have involved the work of meteoric waters. Such waters are also critical in the formation of a newly recognized family of fine-grained silver and gold deposits - the Carlin-type which formed in a hot spring environment.

As a final neptunist note, the manganese nodules now accumulating on the deep sea floor are scavenging immense amounts of copper, nickel, and cobalt from seawater. They have not been considered in this paper since it has not yet been demonstrated that they are ores - that is, that they can be mined at a profit. However, it is likely that manganese nodules will become an important source of these four vital metals sometime in the near future.

Those of my readers who are aficionados of ore deposits will realize that in this brief discourse I have touched on the views of many divided schools of ore genesis-sedimentary, volcanogenic and magmatic; ascensionist, descensionist and lateral secretionist; even neptunist and plutonist and that I have found some merit in each. Surely we can agree with Arthur Holmes, who used a quote from John Woodward to conclude the 1944 Edition of Physical Geology and his then controversial chapter on continental drift. Woodward in 1695 had first used these words to describe ore deposits: "Here," he declared, "is such a vast variety of phenomena and these many of them so delusive that 'tis very hard to escape imposition and mistake."

REFERENCES

(1) Beales, F.W., and Jackson, S.A. 1966. Precipitation of lead-zinc ores in carbonate reservoirs as illustrated by Pine Point ore field, Canada. Inst. Min. Metallurgy Transactions (London), section B. 75: 278-285.

(2) Button, A. 1976. Transvaal and Hamersley basins - review of basin development and mineral deposits. Min. Sci. Engin. (Johannesburg) 8: 262-292.

(3) Cloud, P. 1976. Major features of crustal evolution. Geolog. Soc. South Africa Annex (Johannesburg) 79: 1-33.

(4) Cook, P.J., and McElhinny, M.W. 1979. A reevaluation of the spatial and temporal distribution of sedimentary phosphate in the light of plate tectonics. Econ. Geol. (New Haven) 74: 315-330.

(5) Demaison, G.J., and Moore, G.T. 1980. Anoxic environments and oil source bed genesis. Bull. Am. Asso. Petrol. Geologists (Tulsa) 64: 1179-1209.

(6) Folinsbee, R.E. 1975. Precambrian metallogenetic epochs - atmospheric or centrospheric? In Recent Contributions to Geochemistry and Analytical Chemistry, ed. A.I. Tugarinov, pp. 281-292. Jerusalem: Keter Press Enterprises.

(7) Folinsbee, R.E. 1976. World's view from Alph to Zipf. Geolog. Soc. Am. Bull. (Boulder) 88: 897-907.

(8) Francheteau, J. et al. 1979. Massive deep-sea sulphide ore deposits discovered on the East Pacific Rise. Nature (London) 277: 523-528.

(9) Hallbauer, D.K. 1975. The plant origin of the Witwatersrand 'carbon.' Min. Sci. Engin. (Johannesburg) 7: 111-131.

(10) Hallbauer, D.K. 1980. The paleoenvironment and thermal history of the Witwatersrand fossil placers. Resumés 26e Congrès Géologique International (Paris), vol. 2, Section 10, p. 774.

(11) Haynes, D.W. 1979. Geological technology in mineral resource exploration. Mineral Resources of Australia Third Invitation Symposium (Adelaide), Preprint No. 2, pp. 1-23.

(12) Holland, H.D. 1973. The oceans: a possible source of iron in iron-formations. Econ. Geol. (New Haven) 68: 1169-1172.

(13) Holland, H.D. 1979. Metals in black shales - a reassessment. Econ. Geol. (New Haven) 74: 1676-1680.

(14) Hutchinson, R.W. 1973. Volcanogenic sulfide deposits and their metallogenic significance. Econ. Geol. (New Haven) 68: 1223-1246.

(15) Jacobsen, J.B.E. 1975. Copper deposits in time and space. Min. Sci. Engin. (Johannesburg) 7: 337-371.

(16) North, F.K. 1979. Episodes of source sediment deposition: the episodes in collective overview. J. Petrol. Geol. (Beaconsfield) 2: 199-218.

(17) North, F.K. 1980. Episodes of source sediment deposition: the episodes in individual close-up. J. Petrol. Geol. (Beaconsfield) 2: 323-338

(18) Ohle, E.L. 1980. Some considerations in determining the origin of ore deposits of the Mississippi Valley Type-Part II. Econ. Geol. (New Haven) 75: 161-172.

(19) Page, R.W . 1975. Geochronology of Late Tertiary and Quaternary mineralized intrusive porphyries in the Star Mountains of Papua New Guinea and Irian Java. Econ. Geol. (New Haven) 70: 928-936.

(20) Pretorius, D.A. 1975. The depositional environment of the Witwatersrand goldfields: a chronological review of speculations and observations. Min. Sci. Engin. (Johannesburg) 7: 18-47.

(21) Radke, A.S.; Rye, R.O.; and Dickson, F.W. 1980. Geology and stable isotope studies of the Carlin Gold Deposit, Nevada. Econ. Geol. (New Haven) 75: 641-672.

(22) Rayner, R.A., and Rowlands, N.J. 1980. Stratiform copper in the Late Proterozoic Boorloo delta, South Australia. Mineralium Deposita (Berlin) 15: 139-149.

(23) Rowlands, N.; Drummond, A.J.; Jarvis, D.M.; Warin, O.N.; Kitch, R.B.; and Chuck, R.G. 1978. Gitological aspects of some Adelaidean stratiform copper deposits. Min. Sci. Engin. (Johannesburg) 10: 258-277.

(24) Sangster, D.F. 1972. Precambrian volcanogenic massive sulfide deposits in Canada: a review. Geolog. Surv. Canada (Ottawa), Paper 72-22, pp. 1-43.

(25) Sillitoe, R.H. 1972. A plate tectonic model for the origin of porphyry copper deposits. Econ. Geol. (New Haven) 67: 184-197.

(26) Sheppard, M.F.; Nielsen, R.L.; and Taylor, H.P., Jr. 1971. Hydrogen and oxygen isotope ratios in minerals from porphyry copper deposits. Econ. Geol. (New Haven) 66: 515-542.

(27) Walker, R.R., and Mannard, G.W. 1974. Geology of the Kidd Creek Mine - a progress report. Can. Min. Metallurgical Bull. (Montreal), No. 752, vol. 67: 41-57.

(28) Worthington, J.E.; Kiff, I.T.; Jones, E.M.; and Chapman, P.E. 1980. Applications of the hot springs or fumarolic model in prospecting for lode gold deposits: Min. Engin. (Littleton, CO) 32: 73-79.

(29) Zumberge, J.E.; Sigleo, A.C.; and Nagy, B. 1978. Molecular and elemental analyses of the carbonaceous matter in the gold and uranium bearing Val Reef carbon seams, Witwatersrand sequence. Min. Sci. Engin. (Johannesburg). 10: 223-246.

Mineral Deposits and the Evolution of the Biosphere, eds. H.D. Holland and M. Schidlowski, pp. 237-256. Dahlem Konferenzen, 1982.
Berlin, Heidelberg, New York: Springer-Verlag.

Sedimentary Balance Through Geological Time

A. Lerman
Dept. of Geological Sciences, Northwestern University
Evanston, IL 60201, USA

Abstract. Balance in the sedimentary cycle is maintained by the sum total of weathering, transport, deposition of sediments, and their eventual reincorporation in the continental crust. Biological activity and primarily photosynthesis strongly affect the fluxes and deposition of carbon, sulfur, iron, and phosphorus containing sediments. The limiting-nutrient role of phosphorus is to a variable degree related to mineral solubilities and competition of inorganic sinks (clays, iron oxides) for dissolved phosphate. The efficiency of phosphorus utilization by organisms is greater in land plants (higher carbon/phosphorus ratios) than in marine phytoplankton (lower C/P ratios). Land vegetation is a large reservoir of phosphorus; changes in its size, coupled with decomposition and transport, can constitute significant inputs of phosphorus to oceanic biota. Photosynthetic fixation of carbon and its net storage in sediments control four major fluxes of the global sedimentary cycle: organic matter, limestones (CH_2O, $CaCO_3$), sulfate, and sulfide ($CaSO_4$, FeS_2). Changes in the net deposition rate of organic carbon produce changes in all the other fluxes within the cycle. The magnitude of such changes depends on the nature of reduced iron-mineral sinks for oxygen and the nature of the CO_2 sources replenishing CO_2 consumed in photosynthesis. A negative correlation between the isotopic abundance ratios $^{13}C/^{12}C$ in carbonates and $^{34}S/^{32}S$ in sulfates since the Late Proterozoic (7,14) corroborates the dependence of the sedimentary fluxes on one another.

SEDIMENTARY CYCLE

The Earth's crust, oceans, and atmosphere, interconnected by fluxes of materials, comprise a sedimentary cycle. The major processes or links within the sedimentary cycle are: (a) chemical and physical weathering of crustal rocks; (b) transport of the products of weathering by wind and flowing water to the oceans; (c) deposition of sediments; and (d) reinclusion of sediments in the continental crust by such processes as tectonic plate subduction, remelting, metamorphism, and uplift. The processes (a)-(d) describe schematically a closed cycle, leaving unanswered a question as to what is being added from the Earth's interior, and on what time scales can the sedimentary cycle be considered closed. If the cycle functioned perfectly and all the sediments deposited on the ocean floor during some period of geological time were reincorporated in crustal rocks through subduction and metamorphism, then there would have been little, if any, evidence of the past preserved. Part of the oceanic sediments, in particular those deposited in shallow-water near-shore environments, have escaped the route of subduction and remelting, giving us a glimpse of the past. The most important characteristics of transport within the sedimentary cycle are the roles played in it by fluids (i.e., water and atmosphere) and by living organisms which are jointly responsible for the complex transformation of igneous crustal rocks into sediments.

Transport

The environment of life is primarily the interface between the atmosphere and the continental crust + ocean. Within this zone the major transport mechanisms are rivers, transporting dissolved and solid materials from land to the ocean, and winds, carrying solids from land and, to a lesser extent, oceanic salts from the ocean. Some estimates of the magnitudes of these transport fluxes, compiled from several sources in (10), are as follows:

Wind-transported material	0.6×10^{12} to 1.6×10^{12} kg/yr
Suspended load of rivers	9.0×10^{12} to 18.0×10^{12} kg/yr
Dissolved load of rivers	2.5×10^{12} to 4.0×10^{12} kg/yr

The two major fluxes are the loads carried by rivers. Extreme values of the ratio dissolved/suspended load vary from about 1:2 to 1:7. In times of high continental relief and abundance of exposed crystalline rocks, the fraction of suspended load in rivers can be expected to be higher, although the amounts of dissolved and suspended materials may also be higher than in times of lower rates of erosion. The dissolved load fraction accounts for anywhere from 12 to 30% of the total flux of materials from land to the ocean. Because both the dissolved and suspended materials must eventually exit from ocean water if the composition of ocean water remains constant on a geological time scale, a fairly substantial uncertainty in the rate of delivery of dissolved materials translates into comparable uncertainties in the residence times of individual chemical elements in the ocean. The fact that some elements tend to be transported in solution whereas others are predominantly carried in suspended material is illustrated below for the seven most abundant elements of the Earth's crust and two biologically important elements, phosphorus and sulfur. (Oxygen is excluded from the seven major elements.) The elements in the continental crust, sediments, and river waters are listed as abundances relative to calcium, always taken at the value of 100. The choice of Ca as a normalizing abundance is based on an assumption that virtually all of Ca in marine sediments passes through a solution stage, a choice that is defensible in view of the solubilities of Ca-silicates in crystalline rocks and of carbonate minerals calcite and dolomite. The relative abundances by weight are (from (9)):

	Si	Al	Fe	Ca	Na	Mg	K	P	S
Crust	678	198	136	100	57	56	50	2.5	0.63
Sediments	310	77	48	100	17	31	25	2.0	7.9
Rivers	41	<0.06	4.7	100	28	25	13	0.35	25.

The elements Si, Al, and Fe are primarily transported in suspended solids, as indicated by their low relative abundance in river waters. The differences between the relative abundances of P and S in sediments and rivers indicate that a large fraction of P is brought to the oceans with the suspended materials. Most of sulfur is transported as sulfate ion in solution. The relatively low abundances of P and Si in solution do not prevent them from being vitally important in biological productivity, P in general and Si in one specific group, the diatoms.

Biotic Sedimentary Cycle

From a purely geochemical point of view, life is a cycle of several elements - C, H, O, N, P, S - and it constitutes a part of the global sedimentary cycle. The cycle of biological activity is an integral part of a much bigger sedimentary cycle within which crustal materials are being transformed into sediments. The basic characteristics of the living world that set it apart from the inorganic world are of consequence to the sedimentary cycle only insofar as they alter the magnitudes of the transport fluxes and the composition of sediments. For example, certain rare elements can be concentrated by organisms in their tissues and skeletal parts; biological productivity can be responsible for variable abundances of oxidized and reduced forms of sulfur and carbon in sediments; the emergence of new taxa, such as land plants, and calcareous and siliceous plankton in the ocean, produced new transport mechanisms and new types of sediments.

Perhaps the most pronounced difference between the inorganic geochemical cycle and the life cycle is the dominant role of reduction and oxidation reactions in the living world. A general oxidation-reduction reaction, typical of an autotrophic system, can be written in a form historically attributed to C.B. van Niel (1) as:

$$CO_2 + 2H_2A \rightarrow CH_2O + H_2O + 2A$$

In an ordinary photosynthetic or phototrophic reaction, H_2A stands for H_2O. In chemotrophic reactions, H_2A stands for H_2S or for a variety of organic compounds. The autotrophic reaction produces organic matter (CH_2O in a shorthand notation), some of which is stored in sediments. In green-plant photosynthesis, oxygen is being produced (i.e., $2A = O_2$), with all the important consequences to the weathering reactions and to transport of their products arising from this process.

Nitrogen, phosphorus, and sulfur are not included in the simplistic autotrophic reaction as written above. In view of the critical role of phosphorus in the life cycle, the next section will be devoted to this element in the biosphere and in sediments.

PHOSPHORUS AND THE LIFE CYCLE

Sources and Transport

The background source of dissolved phosphate in waters is probably the calcium phosphate mineral apatite, owing to its widespread abundance as an accessory mineral in granitic rocks. Rocks of granitic composition forming in the late stages of crystallization from melts, known as pegmatites, commonly contain a great variety of phosphate minerals, of exotic and complex composition. Their contribution to the dissolved phosphate flux in rivers, however, must be very small because of their very low abundance in comparison to apatites.

The present-day annual input of phosphorus by rivers to the ocean is small in comparison to the mass of P in ocean water. Even drastic falls or rises in the input flux of P via rivers would have no significant effect on the global average conditions in surface ocean water on a time scale of *ca*. 10^3 years. Internal circulation of the ocean and respiration in the surface water layer supply more phosphorus to the euphotic zone than river inflow. Near-shore ocean waters may be more strongly affected by variations in river input. Land plants may play an important or, perhaps, even a key role in controlling the phosphate input to the ocean. To clarify this point, we should consider the part of the global phosphorus cycle that includes land plants, soils, ocean water, and phytoplankton, as in the following diagram (11):

land plants ⇄ soils —(rivers)→ surface ocean ⇄ phytoplankton

Land plants take up phosphorus from soils and return it in the form of dead humus. Rivers transport phosphate of inorganic origin, weathered from minerals, as well as phosphate from the plant organic matter. Phytoplankton takes P up from surface ocean water and returns part of it through respiration. A small fraction of phytoplankton carries phosphorus to the bottom where it is stored as organic matter in sediments. The present-day estimate of the rate of P uptake or return by land plants from, or to, soils is 30 to 40 times greater than the rate of delivery of dissolved phosphate by rivers to the ocean (11). Because the mass of plants and leaves dying annually is large, relatively small changes in such factors as the amount of vegetation, rates of land erosion, or decomposition of organic matter in soils can change the amount of dissolved phosphate in river input substantially. The emergence of land vegetation during the Paleozoic Era might have created a new delivery mechanism of phosphorus to the oceans. The potential contribution of land plants to the phosphorus load that reaches the ocean can be illustrated by the following numbers: the present-day standing crop of P in land plants is about 20 times that of the oceanic plankton; a 40% increase in the oceanic biomass could be achieved at the expense of 8 to 14% of the land biomass, if the transfer takes place as in the preceding cycle diagram.

Limiting Nutrient

The role of P as a limiting nutrient to phytoplankton is based on the availability of the phosphate-ion in waters. Ultimately, concentrations of dissolved phosphate in lake and ocean waters are controlled by the rates of input from weathered rocks and removal into inorganic and organic sinks. As mentioned previously, the dead organic material stored in sediments is an important organic sink. The competition between inorganic and organic sinks for phosphorus is well demonstrated by the cycle of P in those lakes which undergo seasonal stratification and alternating anoxic and oxidizing conditions. Under oxidizing conditions, $Fe^{(III)}$-hydroxide, precipitating in lakes or transported from land, makes a strongly adsorbing substrate for phosphate. Under reducing conditions, ferric hydroxide is often reduced to a more soluble $Fe^{(II)}$ form, resulting in a net release of iron and phosphorus from lake bottom sediments.

It is conceivable that under globally anoxic conditions the concentration of phosphate in waters was similar to that of the present-day anoxic lakes. With the emergence of oxygen-producing organisms and, more importantly, with the establishment of oxidizing conditions, a new controlling factor - removal of phosphorus into ferric oxides - was created. Such a factor might have become limiting to growth of aquatic organisms in addition to, or instead of, other purely ecological factors, such as population density, food supply, or predation. Among the inorganic sinks, ferric oxides, clays, and apatite are most important. Removal of dissolved phosphate by adsorption on clays, demonstrated repeatedly in soil studies, and precipitation of apatite in sediments are the processes most likely to be responsible for the limiting nutrient role of phosphorus.

If the limiting-nutrient role of phosphorus in water is controlled by inorganic solids (clays, ferric oxides), then we can speculate about evolutionary advantages of a) lithotrophic or chemolithotrophic forms of early life, and b) more efficient mechanisms of P uptake that evolved in organisms. Point a) bears on the hypothesis of life having emerged either in surface waters or on the bottom of shallow seas. Chemolithotrophic forms of life probably found it easier on the bottom because of the abundant substrate material, even if the danger of being covered up by settling sediment was greater there than in surface waters. The present-day occurrences of algal mats in near-shore waters and of phytobenthos in shallow streams suggest that the dangers of silting-up can be avoided successfully. Point b) refers to a more efficient utilization of phosphorus or, in other words, to a greater mass of organic matter produced per unit mass of available P. The carbon/phosphorus atomic ratio in organic materials is one of the possible measures of the efficiency of phosphorus utilization. The ratios C:N:P:S, with P always normalized to unity, are listed (p. 244) for several groups of marine and terrestrial plants. As a whole, land plants are characterized by higher C:P ratios, which probably reflect a lower availability of phosphorus on land and a need in a greater efficiency of phosphorus utilization. Oceanic plankton and algae have lower C:P ratios. The lowest ratio, that for bacteria, is interesting, but it cannot be safely interpreted as a feature inherited from the remote geological past when phosphorus might have been more abundant under anoxic conditions. The C:N:P:S ratios, listed below, are mean values subject to variation of at least by a factor of two in individual cases.

		C	N	P	S
Bacteria (3)		46.5	7.1	1	0.17
Brown algae (3)		318	14	1	2.9
Red algae (3)		244	18	1	7.1
Green Algae (3)		218	10	1	2.4
Oceanic phytoplankton	(2)	137	20	1	1.4
	(5)	332	27	1	2.8
	(12)	106	16	1	1.7
Lichens (3)		1725	47	1	1.4
Ferns (3)		581	23	1	0.5
Gymnosperms (3)		581	35	1	0.5
Angiosperms (3)		1097	52	1	4.9
Land plants	(5)	510	4.2	1	0.76
	(4)	882	9	1	0.6

A hint of possible variations in the phosphate content of the oceans in the geological past is provided by the data on the abundances of organic carbon and phosphate in limestones of different ages, as shown in Fig. 1. There is an overall positive correlation between the amount of organic matter in limestones and their phosphorus content (note that the phosphorus data give the total P content of limestone, not the P content of organic matter in it). The data, taken at their face value, might suggest that there had been periods of higher phosphate concentration in ocean water, as reflected in the higher rates of biological productivity when more organic carbon was produced and stored in sediments, and when more phosphate might have been taken up by $CaCO_3$ of limestones. However, the data in Fig. 1 must be viewed with considerable caution: the positive correlation between the two parameters may reflect a greater abundance of the detrital components in those limestones where the P and C_{org} values are relatively high (J. Veizer, personal communication).

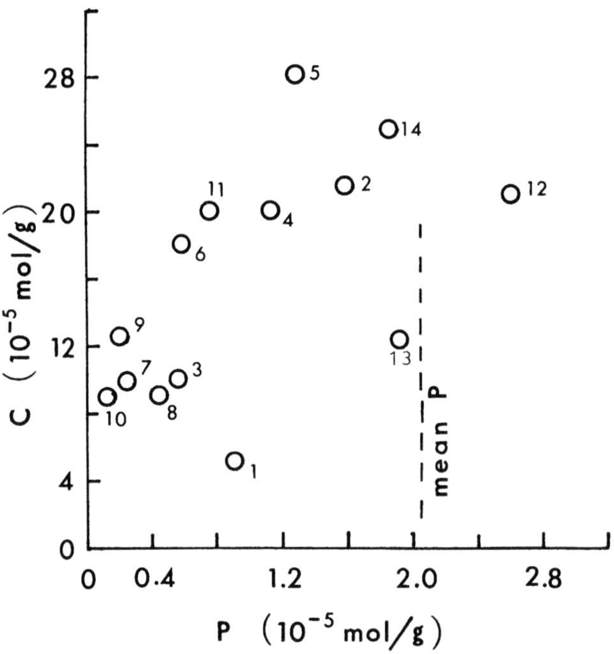

FIG. 1 - Correlation between organic carbon (C) and phosphorus (P) content of limestones, from data in (13). Numbers 1 through 14 indicate ascending age of the samples from 650 million years B.P. to the Tertiary. Mean phosphorus content of limestones (650 ppm) from (3).

This may be a result of a higher abundance of detrital apatite and of detrital clays containing organic matter.

BIOTA AND SEDIMENT FLUXES

Bacterial oxidation and reduction of carbon and sulfur are intimately involved in the formation of C- and S-containing sediments. Figure 2 is a diagrammatic illustration of the sedimentary cycle consisting of the ocean, reduced and oxidized carbon reservoirs (organic matter CH_2O, and limestone and dolomite $CaCO_3$, $MgCO_3$), reduced and oxidized sulfur reservoirs (pyrite in shales FeS_2 and gypsum and anhydrite, denoted $CaSO_4$), and reduced and oxidized iron reservoirs (FeS_2 and Fe_2O_3). This cycle model, worked out by Garrels and Perry (7), postulates

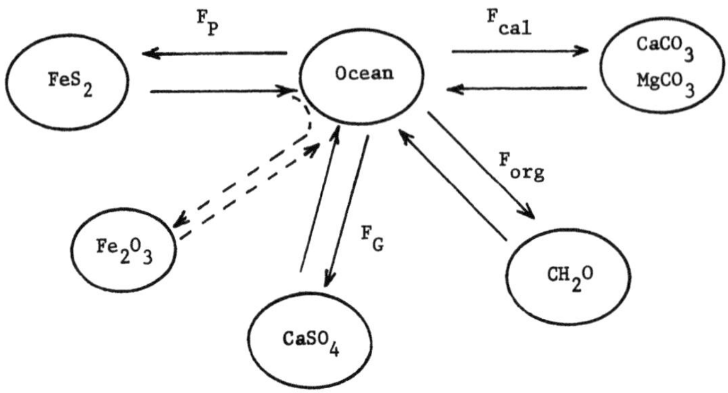

FIG. 2 - Model of the sedimentary cycle in the Phanerozoic, adopted from (7). The sedimentary reservoirs shown are reduced sulfur and iron (FeS_2), sulfate ($CaSO_4$), oxidized iron (Fe_2O_3), reduced carbon (organic matter, CH_2O), and carbonates ($CaCO_3$, $MgCO_3$).

the following. In a closed cycle with ocean of constant composition, a higher rate of biological productivity makes and stores more CH_2O in sediments, which can come only at the expense of a smaller rate of deposition of oxidized carbon in $CaCO_3$. By the same argument, in a finite reservoir of total sulfur, a higher rate of deposition of sulfate in $CaSO_4$ correlates with a lower rate of deposition of reduced sulfur in FeS_2. But, when more organic carbon is being stored in CH_2O, then more oxygen becomes available from reduction of CO_2, and the oxygen can be consumed in oxidation of FeS_2 to SO_4^{2-} and Fe_2O_3. Sulfate is transported to the ocean by streams and its higher input flux (again, in an ocean of constant composition) results in a higher rate of deposition of $CaSO_4$. Thus the coupling between the carbon and sulfur sedimentary cycles requires that during the periods of higher production and storage rates of organic matter, there should be higher rates of deposition of gypsum. In a reverse case, when organic productivity diminishes, less organic matter but more $CaCO_3$ is stored in sediments; the rate of supply of sulfate and deposition of $CaSO_4$ also diminish. In this scheme, changes in the rate of biological productivity can be viewed as driving forces behind the rates of deposition of organic matter and gypsum on a time scale of geological periods.

Despite the effects on the major sedimentary fluxes exerted by changing rates of biological productivity in the past, the effects on the oxygen content of the atmosphere must have been small. The net rate of storage of organic carbon in sediments is between 3.2×10^{12} mol C/yr (6) and 10×10^{12} mol C/yr (8). An equivalent amount of oxygen set free represents a very small fraction of oxygen in the atmosphere (3.8×10^{19} mol O_2) and, in the absence of other oxygen sources, it gives a residence time of 4 to 12 million years for atmospheric oxygen with respect to its net production rate by photosynthesis.

Relationships Between Depositional FLuxes

In a sedimentary cycle depicted in Fig. 2, all the O_2 produced in photosynthesis is consumed in oxidation of FeS_2 to ferric oxides and sulfate in the presence of water. All the CO_2 consumed in photosynthesis for the production of CH_2O is restored by weathering of limestones with a concomitant release of Ca and transfer of Mg from the carbonate (dolomite) to the silicate (clay minerals) pool. It should be stressed that the model in Fig. 2 is only a basic skeleton, useful in a first approximation for a further analysis of the more complicated relationships between the reservoirs and transport fluxes within the sedimentary cycle. Thus, for example, the representation of the carbon cycle as closed ignores the addition of CO_2 to the atmosphere due to metamorphism of limestones (8). When individual reactions for the weathering and deposition processes in Fig. 2 are written out and summed up, the following net balance of fluxes is obtained (7):

$$4FeS_2 + 8CaCO_3 + 7MgCO_3 + 7SiO_2 + 15H_2O \longrightarrow 15CH_2O + 8CaSO_4 + 2Fe_2O_3 + 7MgSiO_3 \qquad [1]$$

The occurrence of the sedimentary reservoirs in reaction [1] - limestone, dolomite, pyrite in shales, oxidized iron, and gypsum - is typical of the Phanerozoic. An alternative source of CO_2 in weathering - siderite ($FeCO_3$) instead of dolomite - has been considered by Garrels and Perry (7) for the Proterozoic. With siderite as a source of CO_2 in the sedimentary cycle model of Fig. 2, reaction [1] assumes a slightly different form (7):

$$3FeS_2 + 7FeCO_3 + 6CaCO_3 + 13H_2O \longrightarrow 13CH_2O + 6CaSO_4 + 5Fe_2O_3 \qquad [2]$$

In [2], the iron sink for oxygen occurs in two mineral forms, as pyrite (FeS_2) and as siderite ($FeCO_3$). From reaction [1] it follows that deposition of 15 molar units of reduced carbon is accompanied by deposition of 8 molar units of oxidized sulfur as sulfate in gypsum. Thus the ratio of the net depositional fluxes of organic carbon (F_{org}) to gypsum is, from [1],

$$F_{org}/F_G = 15/8 = 1.88$$

From the flux balance reaction [2], with siderite in a dual role of an oxygen sink and a CO_2 source, the ratio of the depositional fluxes is

$$F_{org}/F_G = 13/6 = 2.17$$

The difference between the flux ratios in the two models stems from the presence of the different oxygen sinks and different CO_2 sources.

A generalized balance of fluxes in a sedimentary cycle with different $Fe^{(II)}$ sinks for oxygen and different CO_2 sources can be written as:

$$FeS_2 + yFeO + 2CaCO_3 + xCO_2 + (x+2)H_2O \longrightarrow (x+2)CH_2O + 2CaSO_4 + \frac{y+1}{2}Fe_2O_3 \qquad [3]$$

In the above balance FeO stands for any ferrous iron phase, other than FeS_2. CO_2 denotes a source of CO_2 in weathering, in addition to $CaCO_3$, such as dolomite in reaction [1] or siderite in [2] (a case of x = y) or addition of CO_2 from the Earth's interior. In the latter case, the cycle is no longer closed with respect to input from external sources. The stoichiometric key balances in reactions [1]-[3] can be summarized as follows: S in pyrite is balanced by S in gypsum, Ca is conserved between limestone and gypsum, iron is distributed between the $Fe^{(II)}$ and $Fe^{(III)}$ phases, and organic matter in sediments is balanced by C in carbonates and other, if any, CO_2 sources. Pyrite (FeS_2) will be referred to as the main oxygen sink and other $Fe^{(II)}$ minerals as secondary oxygen sinks. A relationship between the stoichiometric coefficients of the fluxes, y and x in [3], follows immediately from the balance equation:

$$y = 4x - 7 \quad [4]$$

Other relationships for the more general case of [3] can be derived similarly. For example, the ratio of the depositional fluxes of organic carbon (F_{org}) and gypsum (F_G) is

$$F_{org}/F_G = (x+2)/2 \quad [5]$$

Balance Eq. [4] is plotted in Fig. 3. The case of no additional oxygen sinks, when only FeS_2 is present, corresponds to y = 0, x = 1.75. In this case, the net flux of CO_2 consumed by photosynthesis in production of CH_2O is balanced either by weathering of limestone and dolomite, as in reaction [1],

$$2CaCO_3 + 1.75MgCO_3,$$

or by weathering of limestone and addition of CO_2 from the Earth's interior, as in reaction [3],

$$2CaCO_3 + 1.75CO_2.$$

The graph in Fig. 3 represents also other conditions within the sedimentary cycle that are discussed below.

At any value of the stoichiometric coefficient y greater than -1, some oxidized iron (Fe_2O_3) is formed at the expense of ferrous iron minerals. The point y = -1 represents a geologically unlikely case of only sulfur, but not iron, being oxidized by photosynthetically produced oxygen. Values of y more negative than -1 ($y < -1$) generally do not represent the sedimentary cycle, as they indicate that ferric iron is being reduced to form ferrous phases. In terms of Eq. [4], the values of y between -1 and -7 correspond to a case of reduction of Fe_2O_3 by reactions with stored organic matter. In this process less CO_2 is required to balance the drain in a steady-state cycle with photosynthesis. For y more negative than -7, the stoichiometric coefficient x becomes also negative, indicating a net production of CO_2 at the expense of organic matter. At the point of y = -15, x+2 = 0, storage of organic carbon in sediments ceases.

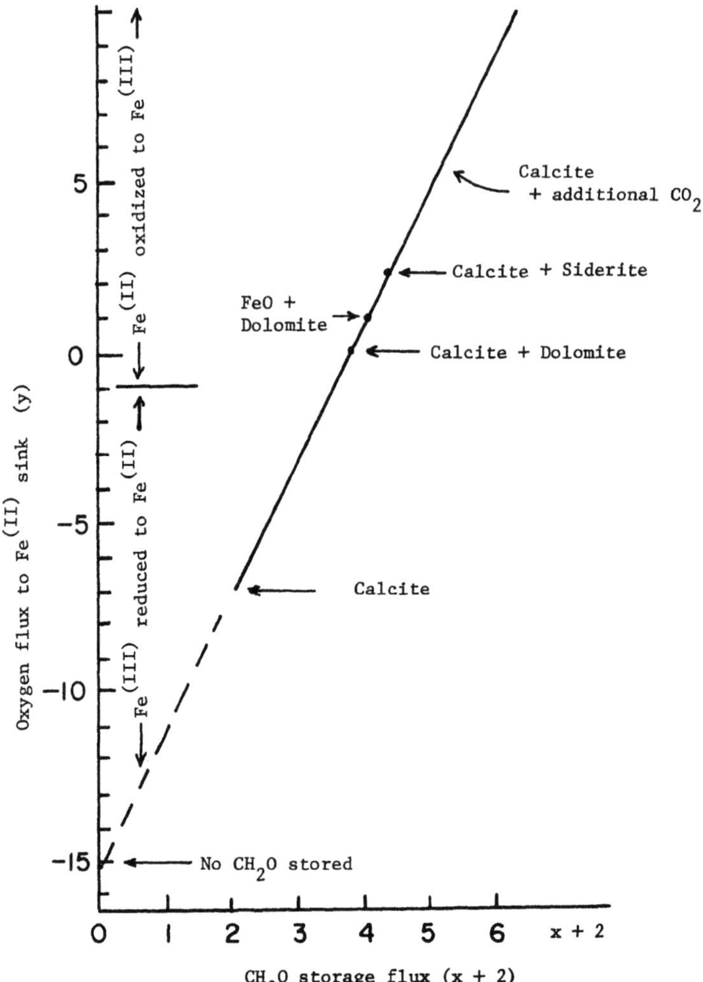

FIG. 3 - Relationship between the flux of oxygen to the $Fe^{(II)}$ sink (stoichiometric coefficient y) and the storage flux of organic carbon in sediments (x+2). Relationship given in Eq. [4].

The more typical sedimentary cycle environments lie on the portion of the graph $y > -1$, $x > 1.5$ or $(x+2) > 3.5$. The portion of the line labeled Calcite + additional CO_2 represents the flux balance reaction [3]; Calcite + Siderite point is given by [2]; Calcite + Dolomite is [1]. The point labeled FeO + Dolomite represents a hypothetical case of no limestone, but only dolomite as a source of CO_2 weathering. The coordinates of this point are y = 1, x = 2, and the flux balance equation follows from [1] and [3] as:

$$FeS_2 + FeO + 2CaMg(CO_3)_2 + 2SiO_2 + 4H_2O \longrightarrow 4CH_2O + 2CaSO_4 + 2MgSiO_3 + Fe_2O_3 \quad [6]$$

The ratio of the depositional fluxes of organic carbon and sulfate, from Eq. [5], lies in the range

F_{org}/F_G = 1.75 to 2.17

Values of the ratio lower than 1.75 represent sedimentary cycles where Fe_2O_3 is reduced, as explained earlier; values higher than 2.17 may represent open cycles with external CO_2 sources.

Depositional Fluxes and Isotopic Evidence

A key evidence to the interdependence of the carbon and sulfur sedimentary cycles in the geological past is the isotopic composition of C in limestones ($CaCO_3$) and of S in evaporitic sulfates ($CaSO_4$), compiled recently by Veizer et al. (14) and reproduced in Fig. 4. The nature of the negative correlation between $\delta^{13}C$ and $\delta^{34}S$ in sediments from Later Proterozoic Time to the Tertiary can be briefly explained as follows (7,14).

Owing to the fact that photosynthetic organisms and sulfur-reducing bacteria utilize preferentially the light isotopes ^{12}C and ^{32}S, C in organic matter and S in FeS_2 are isotopically lighter than in CO_3^{2-} and SO_4^{2-} in the waters, in which they form. A higher rate of organic matter production in the ocean implies removal of more isotopically light C in CH_2O and leaves behind the isotopically heavier carbon in CO_3^{-2} and HCO_3^- in ocean water; $CaCO_3$ precipitating from this water will be characterized by higher values of $\delta^{13}C$. As discussed in a preceding section, times of a higher rate of production or organic matter are also times of a higher rate of sulfate storage in sediments and, correspondingly, a lower rate of FeS_2 formation. At such times, sulfur in the sulfate ion in ocean water would tend to become isotopically lighter and $CaSO_4$ deposited from such waters

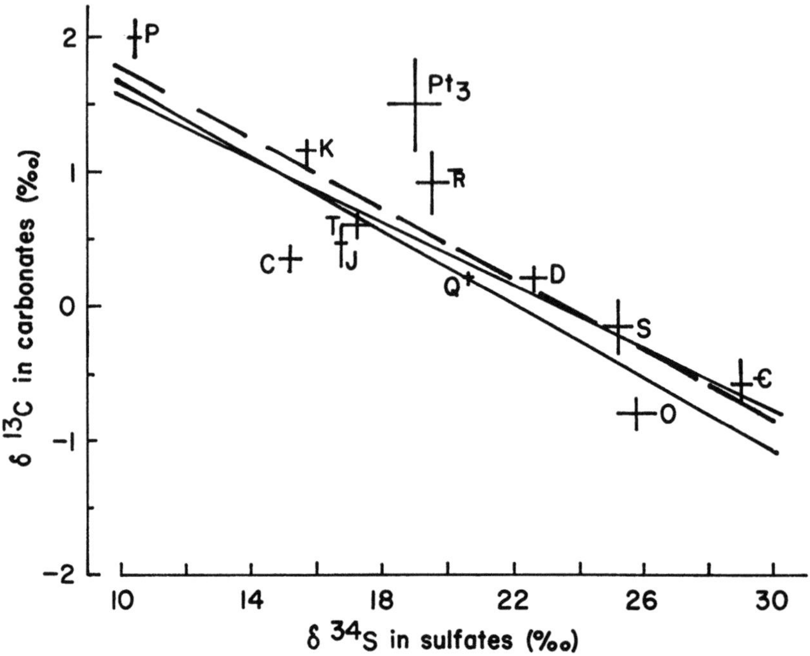

FIG. 4 - Isotopic composition of carbon in limestones and of sulfur in sedimentary sulfates from Later Proterozoic to the Tertiary (14). Dashed line: best fit, Eq. [7]. Solid lines: based on a model, Eqs. [9] and [10].

would be characterized by lower values of $\delta^{34}S$. Thus the periods of higher $\delta^{13}C$ values in limestones would correlate with periods of lower $\delta^{34}S$ values in sulfate evaporites, as Fig. 4 shows. A best fit line to the points is (14)

$$\delta^{13}C = 3.074 - 0.131\ \delta^{34}S \qquad [7]$$

This correlation between the $\delta^{13}C$ values of limestones and $\delta^{34}S$ values of sedimentary sulfates can be also derived in an explicit form from the relationships between the sedimentary fluxes, as in the cycle model of Fig. 2. With reference to the latter, two assumptions may be made: (a) composition of ocean water has remained constant through time (i.e., the masses of sulfur and carbon in the ocean are constant, but not their isotopic composition); (b) the mean isotopic composition of C and S in inflow to the ocean has remained constant through time (the mean isotopic composition of inflow is a weighted sum of the weathering rates and the isotopic δ values of the sulfur and carbon sedimentary reservoirs).

Then a relationship between $\delta^{13}C$ and $\delta^{34}S$ in ocean water, or in other words, in limestone and gypsum forming from ocean water (ignoring small isotopic fractionation factors in precipitation) is given by:

$$\delta^{13}C = \bar{\delta}_C + \frac{\alpha_C(F_{org} + rF_P)}{\Sigma F_C} + \frac{r\alpha_C \Sigma F_S}{\alpha_S \Sigma F_C}(\bar{\delta}_S - \delta^{34}S) \qquad [8]$$

The values of the constant parameters in Eq. [8] that are listed below are taken from as yet unpublished work that also discusses their sources (R.M. Garrels and A. Lerman, Phanerozoic Cycles of Sedimentary Carbon and Sulfur, 1981). The parameters are:

$\bar{\delta}_C = -4.54‰$ and $\bar{\delta}_S = 7.5‰$ are the mean isotopic compositions of C and S in inflow to the ocean;

$\alpha_C = 25‰$ and $\alpha_S = 35‰$ are the isotopic fractionation factors for C in organic matter and S in the reduced sulfur in FeS_2;

$F_{org} = 3.2 \times 10^{12}$ mol C/yr is the net depositional flux of organic matter (CH_2O storage in Fig. 2);

$\Sigma F_C = F_{org} + F_{cal} = 15.7 \times 10^{12}$ mol C/yr is the sum of the depositional fluxes of organic matter and carbonate out of the ocean;

$F_P = 0.48 \times 10^{12}$ mol S/yr is depositional flux of FeS_2;

$\Sigma F_S = F_P + F_G = 1.48 \times 10^{12}$ mol S/yr is the sum of the depositional fluxes of pyrite and gypsum out of the ocean;

$r = F_{org}/F_G$ is the ratio of the storage fluxes of organic carbon and sulfate, as defined in Eq. [5]; mostly $r = 1.75$ to 2.17.

The preceding parameter values and the two bracketing values of the flux ratio $r = F_{org}/F_G$ give two explicit forms of Eq. [8]:

$$\delta^{13}C = 2.77 - 0.12\delta^{34}S \qquad \text{(with } r = 1.75) \qquad [9]$$

$$\delta^{13}C = 3.31 - 0.15\delta^{34}S \qquad \text{(with } r = 2.17) \qquad [10]$$

The straight lines from [9] and [10] drawn through the data in Fig. 4 show that the agreement between the cycle model and the data is very good, despite the fact that the assumptions of a constant ocean and constant mean isotopic composition of inflow to the ocean are undoubtedly oversimplifying the picture.

CONCLUSION

Stated broadly, the sedimentary cycle is driven by physical, chemical, and biological forces. Thermodynamic instability of the minerals formed under conditions different from those of the Earth's surface is the main driving force behind the weathering of crustal rocks and formation of new sedimentary minerals. Biological processes, through reduction and oxidation reactions, modify the fluxes of materials between the major sedimentary reservoirs (e.g., Fig. 2) and in particular those involving the "building blocks" or organic matter - the elements C, H, O, N, P, and S. The important role of P as an element that can often be a limiting factor to biological productivity is probably related to the low solubility of apatite (an abundant phosphate mineral in the crust) and to uptake of dissolved phosphate by inorganic sinks, such as clays and iron hydroxide. It is conceivable that under anoxic conditions and in the absence of $Fe^{(III)}$-oxides, the abundance of phosphorus in waters might have been greater. A measure of the evolutionary efficiency of phosphorus utilization by organisms is provided by the C/P ratios in different plant groups: in marine phytoplankton the C/P ratio is lower than in land plants, possibly reflecting the mode of occurrence and ease of utilization in water as compared to land. The fluxes of oxidized and reduced carbon and sulfur in the sedimentary cycle (reservoirs of $CaCO_3$, CH_2O, $CaSO_4$, and FeS_2 in Fig. 2) are coupled one to another, and this coupling is corroborated by the evidence of the carbon and sulfur isotopes in sediments since the Late Proterozoic (7,14). Quantitatively, the coupling between the carbon and sulfur fluxes, and correlation between the isotopic ratios $^{13}C/^{12}C$ and $^{34}S/^{32}S$ in sediments are accountable for by a model of the sedimentary cycle, based on an ocean of constant mass and on a constant river flux to the ocean that has always averaged the isotopic composition of C and S in sedimentary reservoirs.

Acknowledgement. This work was in part supported by a National Science Foundation grant EAR76-12279. Discussions of this paper with P.J. Cook, G. Eglinton, R.M. Garrels, J. Hoefs, H.D. Holland, I.R. Kaplan, J. Nriagu, and J. Veizer were both stimulating and helpful.

REFERENCES

(1) Blum, H.F. 1955. Time's Arrow and Evolution, 2nd ed. Princeton: University Press.

(2) Bowen, H.J.M. 1966. Trace Elements in Biochemistry. London: Academic Press.

(3) Bowen, H.J.M. 1979. Environmental Chemistry of the Elements. London: Academic Press.

(4) Deevey, E.S., Jr. 1973. Sulfur, nitrogen, and carbon in the biosphere. In Carbon and the Biosphere, eds. G.M. Woodwell and E.V. Pecan. Washington: U.S. Atomic Energy Commission.

(5) Delwiche, C.C., and Likens, G.E. 1977. Biological responses to fossil fuel combustion products. In Global Chemical Cycles and Their Alterations by Man, ed. W. Stumm. Berlin: Dahlem Konferenzen.

(6) Garrels, R.M.; Mackenzie, F.T.; and Hunt, C. 1975. Chemical Cycles and the Global Environment. Los Altos, CA: Kaufmann.

(7) Garrels, R.M., and Perry, E.A., Jr. 1974. Cycling of carbon, sulfur, and oxygen through geologic time. In The Sea, ed. E.D. Goldberg. New York: Wiley-Interscience.

(8) Holland, H.D. 1978. The Chemistry of the Atmosphere and Oceans. New York: Wiley-Interscience.

(9) Lerman, A. 1977. Migrational processes and chemical reactions in interstitial waters. In The Sea, eds. E.D. Goldberg et al. New York: Wiley-Interscience.

(10) Lerman, A. 1979. Geochemical Processes. New York: Wiley-Interscience.

(11) Lerman, A.; Mackenzie, F.T.; and Garrels, R.M. 1975. Modeling of geochemical cycles: phosphorus as an example. Mem. Geol. Soc. Am. 142: 205-218.

(12) Redfield, A.C.; Ketchum, B.H.; and Richards, F.A. 1963. The influence of organisms on the composition of seawater. In The Sea, ed. E.D. Goldberg. New York: Wiley-Interscience.

(13) Ronov, A.B., and Korzina, G.A. 1960. Phosphorus in sedimentary rocks. Geochem. 8: 805-829.

(14) Veizer, J.; Holser, W.T.; and Wilgus, C.K. 1980. Correlation of $^{13}C/^{12}C$ and $^{34}S/^{32}S$ secular variations. Geochim. Cosmochim. Acta 44: 579-588.

Group on Sedimentary Iron Deposits, Evaporites and Phosphorites

Standing, left to right: Peter Cook, Hal James, Malcolm Walter, James Nriagu, Andy Button, Alec Trendall, Tom Brock.
Seated: Hans Eugster, Lynn Margulis, Ken Nealson, Alan Goodwin.

Mineral Deposits and the Evolution of the Biosphere, eds. H.D. Holland and M. Schidlowski, pp. 259-273. Dahlem Konferenzen, 1982.
Berlin, Heidelberg, New York: Springer-Verlag.

Sedimentary Iron Deposits, Evaporites and Phosphorites
State of the Art Report

A. Button, Rapporteur
T. D. Brock, P. J. Cook, H. P. Eugster, A. M. Goodwin,
H. L. James, L. Margulis, K. H. Nealson, J. O. Nriagu,
A. F. Trendall, M. R. Walter

INTRODUCTION

Iron formations, evaporites, and phosphorites are economically important chemical sedimentary rocks developed in basins which span a very large fraction of geologic time. Since they are chemical precipitates, they must reflect the character of the body of water from which they were formed. Our group attempted to determine whether evolutionary changes in the biosphere could be inferred from the stratigraphic record of these sediment types.

The iron-rich sedimentary rocks can be divided into two broad classes: ironstone, which occurs mainly as thin beds of Phanerozoic age; and iron formation, generally chert-banded, mainly of Precambrian age, and representing by far the greatest concentration of iron of sedimentary origin. Our attention was focused almost wholly on the latter group of rocks as potential indices of hydrospheric, atmospheric, and biospheric evolution.

Three principal types of iron formation were distinguished:
1) Algoma type, closely associated with submarine volcanic

strata and considered to be directly related in origin to volcanic exhalations of iron-rich fluids. Chemically and mineralogically these deposits are similar to those of the more extensive Superior or Hamersley type (to be described later), but they tend to be more limited in thickness and areal extent and to contain somewhat greater quantities of base metals. Deposits of this type are particularly abundant in Archean greenstone belts, but they occur in submarine volcanogenic sequences of almost all ages. Their development is considered to be related to periodic tectonic events, rather than to the evolutionary processes to which the attention of this conference is primarily directed. 2) Rapitan type, occurring in several basins of late Precambrian age and at least in part associated stratigraphically with deposits of glacial origin. Several deposits of significant dimensions are known, notably in the Rapitan Group of northwestern Canada and in the Morro du Urucum region of western Brazil and eastern Bolivia. These deposits, apparently somewhat richer in iron than normal iron formations, are not well-known or dated. They may reflect specialized environmental conditions associated with sedimentation at high latitudes, but current information is too sparse to warrant conclusions regarding their significance for the evolution of the earth. 3) Superior and Hamersley types, which represent by far the greatest concentration of iron in the geologic record. The two varieties are distinguished mainly on the basis of internal structures (in Superior-type ore oölitic or granule forms are common), but otherwise they are chemically and mineralogically similar. Both are chert-banded, are commonly hundreds of meters thick and tens or hundreds of thousands of square kilometers in areal extent, and occur in basins of mid-Precambrian age. The largest basins were formed between 2,600 m.y. to 2,000 m.y. This time limitation of the major deposits and their abrupt disappearance from the geologic record some 2,000 m.y. ago are remarkable facts. They suggest that these deposits reflect some major phase or transition in the evolution of the earth's hydrosphere and biosphere. Other significant features of these iron formations are their constancy in gross chemical composition (30% Fe, 45% SiO_2), low content of trace elements, and, in some basins, the remarkable lateral continuity of individual layers and laminae (see James and Trendall, this volume).

There are distinct stratigraphic, sedimentologic, and mineralogic relations between Superior-Hamersley-type iron formations and phosphorites and, to a lesser extent, evaporites. Our discussions centered on these topics. We were particularly concerned with examining the evidence for a biological involvement in the development of these systems and for evolutionary trends in the biosphere as a whole.

SUPERIOR AND HAMERSLEY-TYPE IRON FORMATION

In the past there has been little agreement on the genesis of cherty iron formation of the Superior and Hamersley-types. We were therefore surprised to find essential unanimity within the group regarding some of the most important aspects of the genesis of these sediments. The favored model (Fig.1) involves a Precambrian ocean with an upper oxic layer overlying a much larger volume of anoxic water. Even a relatively low concentration of dissolved ferrous iron in the deep ocean would, because

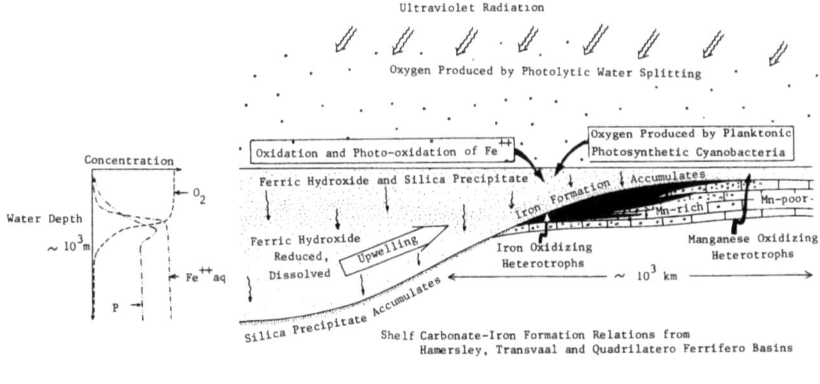

FIG. 1 - Conceptual model for the deposition of mega iron formations (based in part on concepts developed by H.D. Holland and P.E. Cloud, see end of list of references).

of the immense volumes involved, have constituted an enormous reservoir of the metal. The source of iron is no longer considered to be a problem; like sodium in the present ocean it accumulated from terrigenous and submarine sources over a long period of geologic time.

All the organisms capable of forming siliceous skeletons are eukaryotes, either protists (diatoms, radiolarians) or animals (sponges), and do not appear in the fossil record until the Phanerozoic. In the absence of organisms that deplete silica from solution, it is inferred that silica accumulated in the Precambrian oceans, building to concentrations close to saturation with amorphous silica. There could thus have been an essentially constant rain of silica-rich precipitate onto the ocean floor.

Large-scale precipitation of iron in the major iron formations is thought to have begun when deep, iron-rich water gained access to newly-developed shallow continent-margin basins and shelves. The process of transfer to sites of deposition could have been somewhat analogous to the upwelling mechanism that brings deep, cold, phosphorus-rich water onto the shallow shelves of the present ocean; this mechanism is commonly invoked to explain the presence of major phosphate deposits in the stratigraphic record.

There is no consensus regarding the levels of concentration of oxygen or of the mechanism for its generation in the surface layer of the ocean. The most likely explanation is the photosynthetic generation of oxygen by surface scums or mats of cyanobacteria in the photic zone, the upper 200 meters of the ocean. There is direct evidence from the fossil record that such microbial communities thrived even during the Archean. However, we cannot prove that non-biological photooxidation and photolytic dissociation of water vapor were not partly, or even largely, responsible for iron oxidation and precipitation.

There are several alternative (and, for the moment, equally acceptable) ways of explaining the formation of alternating iron-rich and iron-poor microbands in iron formation. There is a strong possibility that the cyclicity represents seasonal changes. Seasonal upwelling of phosphorus (along with iron) may have been followed by a cyanobacterial bloom which gave rise to increases in oxygen production, resulting in a shower of ferric hydroxides. If iron formations were formed under temperate climatic conditions, seasonal variations in oxygen production probably occurred. During summer growing seasons, oxygen release by photoautotrophs may have led to continuous iron and silica precipitation. The uppermost part of the sediment may then have been depleted in iron by subsequent microbial reduction (and consequent solution) during the darker, colder season when heterotrophy dominated over autotrophy.

In addition to providing oxygen for iron oxidation and precipitation, organisms may have played a more direct role in iron deposition. For example, heterotrophic microbes may have concentrated iron hydroxides directly as many "iron bacteria" do today. Such organisms may have thrived at the shallow sediment-water interface, utilizing oxygen and small amounts of organic matter (both provided by photoautotrophs) in the precipitation of iron (and manganese).

It is emphasized that this model is not without problems. For example, could stable shelves and mega iron formations have developed before 2,600 m.y. ago and have been subsequently removed by erosion? Since upwelling onto a shelf requires complementary downwelling (and thus mixing of ocean waters), could the steep oxygen gradient in the ocean have been maintained for millions of years? Given a steep oxygen gradient in the oceans (Fig. 1), it is logical that there would also have been a steep gradient in the concentration of dissolved ferrous iron, which would be very low near the surface. Could this concentration gradient alone have caused a continuous migration of dissolved iron to the upper level of the ocean? Despite these and other questions, it is thought that the model described in the preceding paragraphs explains more facts than other models and provides an excellent framework within which more data can be gathered to confirm, modify, or reject the concept.

EVAPORITES

The mechanisms of evaporite formation are fairly well understood. Precambrian evaporites provide important constraints on the physical and chemical parameters of the ancient oceans. Unfortunately, bedded evaporites have a relatively short residence time in the stratigraphic record. In rocks as old as the Precambrian, original evaporites are almost always removed by solution, so that we usually have to rely on the presence of crystal casts of minerals such as gypsum and halite.

In relatively low salinity water, gypsum (but not anhydrite) is stable as a precipitate at water temperatures of up to 57°C. Where salinities approach the halite saturation level, gypsum is unstable above 25°C. Halite and gypsum casts together provide constraints on water temperature and can be useful paleoclimatologic indicators.

There are several connections between evaporite development and biologic processes. For example, if gypsum is present in an ancient basin, sulfate concentrations are greater than the 1 or 2 ppm needed to support a sulfate reducing bacterial system. There are other more direct links between biologic processes and the formation of evaporites. For example, Na or Na-Ca carbonates can be precipitated from alkaline brines by the addition of CO_2 given off by fermenting organic substances in the basin of deposition.

At present, the oldest known evaporites are the barite-filled gypsum casts from the 3,500 m.y. old Warrawoona Group in Western Australia (1). Other Precambrian evaporite occurrences are being found by careful studies of sedimentary successions; but much more geologic data, gathered on a global basis, are needed before we can use evaporites with confidence to reconstruct the geochemical evolution of the oceans. The development of geochemical constraints for the Precambrian oceans is sorely needed to advance our knowledge of the evolutionary changes in the interdependence of organisms and their chemical environment. A more pragmatic goal of such studies might relate to ore deposit genesis. Evaporitic brines are known to play a very important part in the transport of metals such as Cu and Ag.

PHOSPHORITES

Phosphorus, an essential component of nucleic acids, is required by all living systems. It can be inferred to have been cycled between seawater and organisms for as long as life has existed on earth. With some qualifications (e.g., detrital apatite grains in a sediment), the spatial and temporal distribution of phosphorus in the organic carbon fraction of marine sediments might be expected to provide a continuous record of the activity of organisms throughout the stratigraphic record. In this regard, phosphorus will probably prove to be more useful than elements such as H, C, O, and N, which are lost in varying amounts as sedimentary organic matter matures after it is buried.

The distribution of phosphorite deposits in Phanerozoic sediments is known to be related to their paleolatitude and to the configuration of the ancient continental shelves (3,12). Through the Phanerozoic, sites of upwelling and of phosphate deposition have mainly been at low latitudes (30 degrees north and south of the equator). The distribution of the element may well be found to be an excellent paleogeographic indicator in Precambrian basins of all ages.

Phosphorus is a potentially useful indicator of biologic activity and of paleogeography (see above); it is therefore surprising that there is little or no systematically collected data on the distribution of this element through the first 85 percent of the time for which a stratigraphic record exists.

The oldest known phosphorites are apatite-rich mesobands (a few centimeters thick) in 2,600- to 2,000-m.y.-old iron formations. The distribution of phosphorus in these rocks is relatively well studied only because the element is an undesirable constituent of iron ores. Phosphatic stromatolites are known from the 1,200- to 1,600-m.y.-old Aravalli Group of India. A systematic search will undoubtedly result in the discovery of many more phosphorites in Proterozoic (and possibly even Archean) basins.

Phosphorites are somewhat more abundant in basins younger than about 900 m.y. (4). However, the most striking increase in the abundance of phosphorites occurs at about the Precambrian-Cambrian boundary (3,4,13). It seems reasonable to infer that the change is linked to evolutionary advances in the biota, such as the development of phosphatic exoskeletons. The study of the distribution of phosphorus in the organic carbon fraction of Precambrian sediments appears to be a very fruitful field for research. It offers promise of providing a recording of the intensity of biological activity through time and of assisting in the paleogeographical reconstruction of marine basins through the Precambrian.

OPEN QUESTIONS AND AVENUES FOR NEW RESEARCH

The conference revealed how much information is needed in order to understand the origin and interrelationships of iron formations, evaporites, and phosphorites. The following list includes some of the fields in which data are sorely missed and some of the questions which seemed particularly pertinent.

Temporal Distribution of Iron Formation, Evaporites, and Phosphorites

There is a great need for continued improvement in the precision of age measurements in basins containing the chemical sediments in question. If ages could be established well enough to allow the calculation of deposition rates, narrower constraints could be placed on geochemical models of deposition. It may be possible to decide, for example, whether microbands in Hamersley-type iron formations are seasonal layers.

Physical Aspects of the Depositional Environment

The character of the Precambrian oceanic crust is largely unknown. Given the inferred high heat flow, it has been calculated that cycling of ocean water through this crust resulted in a much higher flux of iron, manganese, and silica from sea-floor hot springs than is presently the case. Can this model be independently confirmed and, if so, is this an adequate source of iron and silica from the mega iron formations?

There is a great need for paleomagnetic basinal studies to establish paleolatitude and the configuration of the supercontinents. If determined, the paleolatitude of Precambrian basins would allow us to interpret the significance of (for example) evaporation and glaciation in the deposition of the chemical sediments in question. We might also be able to confirm that some of the layering in iron formation is indeed due to seasonal climatic cycles.

Given available data, what can we infer about the tectonic setting of basins containing Precambrian evaporites and phosphorites? If the answer is 'nothing,' what information should be collected in an attempt to answer questions concerning the tectonic setting of evaporitic and phosphogenic basins?

Can we predict the degree to which photooxidation of divalent iron might have contributed ferric precipitates to basins of iron formation deposition? Could photolytically-produced oxygen account for a significant part of the oxygen in iron oxides fixed in iron formations?

Chemical and Biological Aspects of the Ocean/Atmosphere System

The favored model for iron formation deposition suggests a large reservoir of reducing oceanic water and a weakly oxidizing atmosphere. Can these assumptions be independently verified by study of, for example, subaerial and submarine paleoweathering profiles and products?

The model for iron formation deposition suggests that separation of iron from manganese is due to preferential removal of iron (by oxidation and precipitation) from ocean water on the outer shelf. This model is consistent with geologic relations in the Hamersley Basin (W. Australia) and the Transvaal Basin (South Africa). However, other major basins of iron formation deposition (Lake Superior, for example) apparently do not contain manganese-rich carbonates. What is the fate of the Mn that we infer must have accompanied divalent iron in the deep ocean water that gave rise to the Superior iron formations?

What is the significance of the remarkable constancy in the gross chemical composition of cherty iron formation through geologic time? Are trace elements in iron formation as constant and as predictable as the major elements?

Where could we look for contemporary environments which might be biologic analogues of the iron formation basins? Would experimental bacterial modeling be a useful approach to the question of biological involvement in iron formation deposition? Could there have been precipitation of iron by organisms without known counterparts on the contemporary earth?

Is it possible that microbanding (see James and Trendall, this volume) in iron formation and other chemical sediments is biologically controlled, or can microbands be explained purely by physicochemical controls such as seasonally-regulated temperature changes?

Is the precipitation of chert in iron formation and Precambrian phosphorites purely a function of silica saturation in the ocean? What direct or indirect role might organisms have played in silica deposition? How much of the chert in the basins in question might have been formed from chert precursors such as the magadiite (hydrated sodium silicate) now forming in evaporative alkaline lakes of East Africa?

Isotopic work suggests differences in the oxygen isotopic composition of different iron formation microbands (M.R. Walter, personal communication, 1980). Does this reflect seasonal changes in seawater properties or microbial communites or both?

What microbial systems should we study to allow us to start asking the right questions regarding a possible biological control on the deposition of iron formation? What are the most important Fe and Mn reducing and oxidizing bacteria, and in what ranges of pH and Eh do they influence the solution or precipitation of the metals in question?

Could there be Precambrian counterparts to presently known magnetite-bearing bacteria and, if so, could they have played a significant part in iron formation deposition? Could bacterial (or, for that matter, inorganic) reduction have caused solution of the iron oxide fraction in layers composed of a mixture of chert and iron oxide? If so, microbanding could be a solutional rather than a depositional phenomenon. What new insights would this lend to our interpretation of microbands? The microbiology of diagenetic environments is very poorly known. Can we gain a new understanding of iron formation diagenesis by looking at bacterial influences on recently deposited iron-rich sediments?

What do we not know about the phosphorus cycles in the ocean? The geochemical cycle of phosphorus in the present ocean should be better quantified to allow us to interpret the stratigraphic record. Data on the C,N, and P content in the organic fractions of the Precambrian sediments should be collected. We need to study the trace element distribution patterns in phosphorites to assist us in reconstructing the chemistry of the ocean in times past. Can we be sure that bacterial or algal fixation of phosphorus was not responsible for the formation of phosphatic stromatolites? Far better documentation of Recent phosphorites on the present day sea floor is essential in order to really understand the nature and significance of ancient phosphorites.

Mineralogy and Geochemistry of the Sediments

Many iron formations are enriched in phosphorus. What is the phosphorus distribution in other Precambrian sedimentary rocks? What other mineralogic and geochemical peculiarities are there in iron formation associated sediments and in Precambrian sediments in general? Is it possible to monitor the C, N, and P levels in the organic fraction of Precambrian sediments of all ages? Is such an approach likely to yield useful information on the evolution of life and of the atmosphere-ocean system?

We know relatively little about the nature of precipitates or the kinetics of precipitation from reducing high-iron, high-silica waters. We believe that experimental work along these lines should be done and that it may yield important insights into depositional processes of iron formation.

How similar are the late Precambrian (Rapitan-type) iron formations to those of the older basins? Were conditions for their formation significantly different from those for early Proterozoic iron formation? In the case of the late Precambrian iron formations, how do we develop a reservoir of reduced iron in an ocean in contact with an atmosphere containing near-present oxygen levels?

At present, our ability to interpret evaporite-related structures is limited. Can we develop criteria that will allow us to distinguish diagenetic from syngenetic crystal casts? Can we develop new criteria of crystal cast morphology that will allow reconstruction of the chemical and physical conditions in the depositing medium? What other sedimentary structures (such as mudcracks and caliche nodules) are there that might allow us to more fully determine the character of the evaporative system?

SUMMARY

This meeting has resulted in an exchange of views and information between scientists with varied backgrounds and specialties. New insights and ideas have been gained in several areas. For example, few geologists working with Precambrian sediments were aware of the implications of the above-normal levels of phosphorus in iron formation. In the future the concepts of upwelling and of possible biological fixation of the element will have to be considered when discussing iron formation genesis. There is also motivation now to collect new data on phosphorus to determine basinal paleogeography.

One of the most intriguing new ideas is the possibility that microbands need not necessarily represent depositional features but could be due to bacterial reduction of iron and hence its solution and removal from a chert-iron oxide mixture. This idea stimulates reevaluation of the interpretation of microbanding and leads to other questions such as: Can the iron solution and removal be achieved abiologically? What other diagenetic changes are biologically controlled?

The definition of evolutionary changes has proved to be an elusive goal. Some of the geological events we identified happened only once but we hesitate to call them evolutionary. For example, the iron formation 'peak' (2,600 to 2,000 m.y.) appears to represent the serendipitous development of continental shelves at a time when the lower levels of the ocean were anaerobic and charged with ferrous iron. Other changes may be truly evolutionary; the peak in phosphogenesis near the Precambrian-Phanerozoic boundary almost certainly coincides with a physiological advance. There could be older peaks in phosphogenesis which represent even earlier evolutionary advances, but we simply do not have the data needed to demonstrate such changes.

As can be seen from this report, the discussions produced some consensus, a few answers, and many questions. We feel encouraged; asking the right questions will orient our research in new profitable directions.

REFERENCES

(1) Barley, M.E.; Dunlop, J.S.R.; Glover, J.E.; and Groves, D.E. 1979. Sedimentary evidence for an Archean shallow-water volcanic-sedimentary facies, eastern Pilbara Block, Western Australia. Earth Planet. Sci. Lett. 43: 74-84.

(2) Cloud, P.E. 1976. Beginnings of biospheric evolution and their biochemical consequences. Paleobiol. 2: 351-387.

(3) Cook, P.J. 1976. Sedimentary phosphate deposits. In Handbook of Stratabound and Stratiform Ores, ed. K.H. Wolf, pp. 505-535. New York: Elsevier Press.

(4) Cook, P.J., and McElhinny, M.W. 1979. A reevaluation of the spatial and temporal distribution of sedimentary phosphate deposits in the light of plate tectonics. Econ. Geol. 74: 315-330.

(5) Drever, J.I. 1974. Geochemical model for the origin of Precambrian banded iron formations. Geol. Soc. Am. Bull. 85: 1099-1106.

(6) Goodwin, A.; Monster, J.; and Thode, H.G. 1976. Carbon and sulfur isotope abundances in Archean iron formations and Early Precambrian life. Econ. Geol. 71: 870-891.

(7) Gulbrandsen, R.A. 1969. Physical and chemical factors in the formation of marine phosphorites. Econ. Geol. 64: 365-382.

(8) James, H.L. 1954. Sedimentary facies of iron formations. Econ. Geol. 49: 235-293.

(9) Kuztnetsov, S.I. 1970. The Microflora of Lakes and Its Geochemical Activity. Austin, TX: University of Texas Press.

(10) Lundgren, D.G., and Dean, W. 1979. Biogeochemistry of iron. In Biogeochemical Cycling of Mineral Forming Elements, eds. P.A. Trudinger and D.J. Swaine, pp. 202-211. Amsterdam: Elsevier Press.

(11) Perfil'ev, B.V.; Gabe, D.R.; Gal'perina, A.M.; Rabinovich, V.A.; Sapotniskii, A.A.; Sherman, E.E.; and Troshanov, E.P. 1965. Applied Capillary Microscopy. New York: Consultants Bureau.

(12) Sheldon, R.P. 1964. Palaeolatitudinal and palaeogeographic distribution of phosphate. U.S. Geological Survey, Professional Paper No. 501-C: C106-C113.

(13) Sheldon, R.P. 1981. Ancient marine phosphorites. Ann. Rev. Earth Planet. Sci. 9: in press.

(14) Trendall, A.F. 1972. Revolution in earth history. J. Geol. Soc. Australia 19: 287-311.

SOURCE BOOK

Economic Geology. 1973. Vol. 68, No. 7, Precambrian Iron-Formations of the World.

Group on Stratified Sulfide Deposits

Standing, left to right: Udo Haack, Hans Trüper, Yehuda Cohen, Phil Trudinger, Rolf Hallberg, Kurt von Gehlen.
Seated: Heimo Nielsen, Ian Kaplan, Neil Williams, Don Sangster.

Mineral Deposits and the Evolution of the Biosphere, eds. H.D. Holland and M. Schidlowski, pp. 275-286. Dahlem Konferenzen, 1982.
Berlin, Heidelberg, New York: Springer-Verlag.

Stratified Sulfide Deposits
State of the Art Report

N. Williams, Rapporteur
Y. Cohen, U. Haack, R. O. Hallberg, I. R. Kaplan, H. Nielsen,
D. F. Sangster, P. A. Trudinger, H. G. Trüper, K. von Gehlen

INTRODUCTION

The geology of stratified sulfide deposits suggests strongly that they formed under low temperature conditions at or close to the sediment surface (P.A. Trudinger and N. Williams, this volume). Today, in such environments, iron sulfides frequently form as a result of microbiological sulfate reduction (8), and it has been widely postulated ((18), Table X) that this process was also important in the formation of many of the ancient stratified sulfide deposits. If this postulate is true, one may be able to evaluate the environmental conditions in terms of the oceanic and atmospheric chemistry at the time of formation of the deposits.

To assess whether stratified sulfide deposits are important in this regard, we approached the subject from several directions but were mainly guided by the principle of uniformitarianism: "the present is the key to the past." We first reviewed present-day microbiological processes involved in the sulfur cycle; we next discussed the formation of sulfide minerals in modern sedimentary environments (both at and below the sediment surface); and finally we examined ancient stratified sulfide deposits and discussed how they might be related to modern sulfide-forming environments. Discussions centered on several questions; the results of our deliberations are summarized below.

HOW DO MICROBIAL PROCESSES PRODUCE SULFIDES IN MODERN ENVIRONMENTS?

The metabolic reactions and organisms comprising sulfureta are outlined by H.G. Trüper (this volume). A sulfuretum involves a continual cycling of sulfur between reduced (generally H_2S) and oxidized states (generally sulfate) and it was recognized that acculmulation of a particular sulfur species represents a departure from a steady-state condition.

The most important sulfide-producing process is dissimilatory sulfate reduction, in which organic compounds (or H_2) are oxidized and sulfate is simultaneously reduced to hydrogen sulfide. It was noted that the results of recent studies of sulfate-reducing bacteria have modified our notions of the significance of sulfate reduction in the mineralization of organic matter in the natural environment.

The "classical" dissimilatory sulfate reducers carry out partial oxidation of a few simple organic molecules (e.g., ethanol, lactate, and pyruvate) to acetate and carbon dioxide. Bacteria have now been described which oxidize acetate (and/or some higher fatty acids) to CO_2 and water; this demonstrates that the sulfate-reducers, as a group, are capable of competing with methanogenic bacteria in catalyzing the terminal stages of the mineralization of organic matter. Organic compounds now known to be metabolizeable by sulfate reducers include fatty acids up to C_{18} and benzoate. These observations may help account for the fact that in some shallow-water marine sediments sulfate reduction can account for 50% or more of the degradation of the organic matter produced in the sediments (7). It was stressed, however, that information regarding the ecology of the newly-discovered genera and species of sulfate reducers is lacking, and that the quantitative significance of their metabolic reactions in natural environments remains to be assessed.

As yet, no sulfate reducers have been described which can oxidize amino acids, nucleotides, sugars, and the complex macromolecules which comprise the bulk of the primary biomass. A systematic search for such organisms might well prove rewarding. However, we recognize that oxidative and fermentative organisms are able to degrade complex organic matter partially to compounds that can be used by known sulfate reducers.

The possibility that bacterial sulfate reduction could be coupled to methane oxidation was considered; today there is frequently a spatial separation of methane and sulfide in sediments; methane in the primitive anoxic atmosphere may have provided an energy source for early sulfate-reducers. However, so far there is no evidence that sulfate-reducers per se can metabolize significant amounts of methane, but methane oxidation in anaerobic sediments has been demonstrated, and methanogenic bacteria have been shown to catalyze methane oxidation to acetate to a small extent. The prospect thus exists for a coupled reaction leading to methane oxidation at the expense of sulfate reduction:

1) methanogens: $2CH_4 + 2H_2O \rightarrow CH_3COOH + 4H_2$

2) sulfate reducers: $CH_3COOH + 4H_2 + 2SO_4^{2-} \rightarrow 2CO_2 + 6H_2O + 2S^{2-}$

We recognized that no definitive statements can be made regarding the early evolution of organisms of sulfureta although, on energetic and mechanistic grounds, it seemed reasonable to propose that dissimilatory sulfate reduction evolved from bacterial sulfur-linked anaerobic photosynthesis (H.G. Trüper, this volume). It was also considered plausible that sulfate reduction was preceded by enzymatically simpler and kinetically more facile reductions of partially oxidized forms of sulfur such as elemental sulfur, thiosulfate, or sulfite. It is distinctly possible that primitive sulfureta were based on the recycling of elemental sulfur and H_2S, a process which is accomplished at the present time for example, by Desulfomonas in conjunction with Chlorobiaceae and certain cyanobacteria. The proposed appearance of oxidized sulfur species in the presumed anoxic Archaean-Early Proterozoic environments (S.L. Miller, this volume) presents few problems. Sulfur dioxide could arise from volcanic emanations, or from disproportionation reactions (particularly under photochemical influence) between sulfur dioxide (or sulfite) and sulfide which would produce a variety of oxidation states including sulfate.

The recent discovery that many cyanobacteria can carry out anoxygenic bacterial-type photosynthesis as well as oxygenic photosynthesis is of evolutionary significance and has important implications for the interpretation of palaeontological evidence in ancient rocks. The presence of cyanobacterial fossils can no longer be taken as unequivocal evidence of biological oxygen production. Moreover, stromatolites are no longer necessarily indicative of oxic environments since similar structures are produced by cyanobacteria in present-day anaerobic environments, for example, in Solar Lake, Sinai.

Conditions suitable for biogenic sulfide formation are not uncommon today. Microbially produced sulfides occur in many environments, including stratified basins like the Black Sea, where anoxic conditions exist above the sediment-water interface, and in a range of less restricted marine and non-marine environments, where the rate of supply of detrital organic matter is such that anoxic conditions can be maintained below the sediment-water interface. Highly reducing and less restrictive environments where anoxic conditions can be maintained form in areas of extensive algal mat development, (see P.A. Trudinger and N. Williams, this volume).

The most abundant sulfide mineral in these environments is pyrite. The synthesis of pyrite is complex, involving a series of reactions. Iron monosulfides, such as mackinawite (FeS), may form first, and sulfur-rich phases, such as greigite (Fe_3S_4) may form later. They are in time altered to the end member, pyrite, by reaction with polysulfides and/or elemental sulfur. The controls on pyrite formation in nature are difficult to determine: there are no known effects of biological activity although, under some circumstances, photosynthetic or chemosynthetic sulfide oxidation and/or dissimilatory sulfur reduction may control the level of sulfur and polysulfide. Preliminary laboratory and field studies suggest that pyrite formation is influenced by a variety of physicochemical parameters including temperature, the availability of elemental sulfur, and pH, which in turn controls the sulfide-elemental sulfur-polysulfide equilibria.

As pointed out by P.A. Trudinger and N. Williams (this volume), there are no known modern examples of biologically-produced sulfide accumulations which could be considered as incipient ore bodies. Metal sulfide levels in normal marine sediments are typically 1-2% by dry weight and rarely exceed 5%, and by far the most abundant of these are iron sulfides. These low levels often represent only a few per cent of the total sulfide produced biologically and reflect the relative paucity of reactable metals (and the almost complete absence of metals other than iron) in normal marine environments. Quite clearly, therefore, the formation of a sulfide ore body from biogenic sulfide requires a special set of circumstances in which a quantity of metals is available (see below).

WHAT CRITERIA CAN BE USED TO IDENTIFY SULFIDE MINERALS FORMED AS A CONSEQUENCE OF MICROBIAL SULFATE REDUCTION?

In discussing the formation of pyrite as a consequence of microbial sulfate reduction, we paid particular attention to the criteria that can be used to recognize this sulfide mineral-forming mechanism in ancient sediments. Frequently, sulfur isotopic ratios have been used to distinguish between biogenic and abiogenic sources of sulfide: a variable and isotopically light sulfur isotopic composition (relative to contemporaneous seawater sulfate) is taken as diagnostic of a biological mechanism (12). Studies of modern environments however, indicate that many different patterns of sulfur isotopic composition can be produced by biogenic processes, including compositions which, by traditional interpretations, would be considered abiological; for example, in lacustrine situations $\delta^{34}S$ values may be relatively constant and close to 0 ‰ (2). The situation is further complicated by studies demonstrating how "biogenic" $\delta^{34}S$ patterns in sulfides can also be produced abiologically (10).

In attempting to resolve the problem of recognizing sulfide minerals derived from biogenic H_2S, there was a consensus of opinion that more than one criterion should be used. We discussed various possible criteria and concluded that the origin of sulfide minerals could be more accurately identified if $\delta^{34}S$ data were coupled with other data such as:

a) variations in the ratio of trace-metal concentrations, such as the Cu:Zn ratio in the host rocks. Such ratios appear from laboratory and field studies to be sensitive to changes in redox boundaries within modern sediments in which biogenic sulfate-reduction is occurring (5);

b) strong positive correlations between sulfide-S and organic-C in the host rocks. In modern environments such correlations are frequently characteristic of sediments containing pyrite that formed as a consequence of microbial sulfate reduction (1,4). However, a lack of a correlation between the sulfide-S and organic-C content of sediments does not preclude the operation of biogenic processes;

c) a framboidal habit for pyrite in the host rocks. This habit is common for pyrite formed as a consequence of microbial sulfate reduction in modern environments (17), but as a criterion for confirming the process it must be applied with caution because framboids have also been synthesized inorganically in the laboratory (8); and

d) the identification, in ancient sediments, of possible precursor iron-sulfide phases to pyrite (e.g., mackinawite and greigite).

Because of the important role of organic matter in the formation of biogenic sulfide, it was widely felt that a search for organic compounds diagnostic of bacterial sulfate reduction in sediments might yield very valuable results.

ARE STRATIFIED SULFIDE DEPOSITS FORMING IN THE MODERN ENVIRONMENT?

The answer to this question is clearly yes. The best example of modern stratified sulfide deposits are the Red Sea brine pools (3,14): the areas surrounding the "black smokers" at 21°N on the East Pacific Rise are minor by comparison ((6,16); see also R.E. Folinsbee, and P.A. Trudinger and N. Williams, this volume). It appears from available data that in both areas sulfide formation is abiogenic and related to exhalative processes involving the debouching of hot fluids containing reduced sulfur and metals onto the sea floor.

We next asked the question: could stratified sulfide deposits also form biologically? As discussed earlier, the formation of such deposits would appear to require concentrations of metals far in excess of those found in normal sedimentary environments. It was agreed that should a supply of metals be available (such as from some type of metal-rich fluid), sufficient sulfide could be generated biologically to form a massive, layered sulfide deposit (11,19).

The most probable environments for the formation of the required metal-rich fluids would, we felt, be subsurface environments where metals are extracted during fluid-rock interaction (see (15)). Examples of modern sedimentary environments where such fluids might encounter abundant biogenic sulfide are the anoxic lakes of the East African Rift System (P.A. Trudinger and N. Williams, this volume) and the Guaymas Basin, Gulf of California (9).

Biological processes, although apparently not involved in the formation of the 21°N "black-smoker" sulfides, may be important in their destruction. Sulfide-oxidizing bacteria exist in the vicinity of the "smokers," and the oxidation of sulfides by these bacteria may be responsible for the maintenance of the spectacular fauna around the "smokers" (16). That the interaction between seawater and hot metal-rich fluids can produce local biologic communities is intriguing and may provide an explanation for the abundance of carbonaceous matter in and around many ancient stratified sulfide deposits.

BY WHAT MECHANISMS DID ANCIENT STRATIFIED SULFIDE DEPOSITS FORM, PARTICULARLY THOSE OF PRECAMBRIAN AGE, AND WHAT IS THEIR USEFULNESS IN ELUCIDATINC BIOSPHERIC EVOLUTIONARY PROCESSES?

Reviews of the types of stratified sulfide deposits occurring in the Precambrain reveal that the most important types are the massive sulfide deposits (generally rich in Zn, Cu, Pb, and Ag) that occur in both volcanic and nonvolcanic terrains. The deposits are typically sulfide-rich ($\geq$ 15 wt.% S), although the sulfur content of adjacent rocks may be as low as 1-2 wt.%. Economically important, but less numerous, are the stratabound copper deposits such as those of the Zambian Cu-belt and at White Pine, USA.

After examining many features of the sulfide-rich types of deposit, including geological settings, sulfur-isotopic patterns, and organic carbon contents, we found great difficulty in linking their formation to biological processes. On the other hand, a good argument can be made for the operation of nonbiological ore-forming processes involving the introduction of metal- and sulfide-rich fluids into depositional environments.

It was noted, however, that in addition to non-biological sulfides, some deposits (e.g., those at McArthur River, Australia - see P.A. Trudinger and N. Williams, this volume) may also contain sulfides that formed as a consequence of microbial sulfate reduction. We recognized that there is strong evidence that most of the base-metal sulfide in the stratabound copper deposits, which contain only a few weight percent sulfide, had a biological source. There is also evidence that the sulfides in these deposits were not formed in a near-surface environment, but were deposited by later diagenetic processes which may have involved reactions between earlier-formed biogenic sulfides (e.g., pyrite) and metalliferous fluids. In these deposits deposition of the base-metal sulfides is therefore not a direct result of biological processes operating at the time of deposition of the host sediments. This separation in time between host-rock sedimentation and the formation of metal sulfides limits the use of stratabound copper deposits in studying biospheric evolution.

Since we see little evidence for the presence of biogenic sulfides in the other ancient stratified sulfide deposits, we are forced to conclude that they also offer little direct evidence regarding the evolution of microbial sulfate-reducing processes. We do, however, recognize the dangers of over-generalization and stress that no deposit should be dismissed out of hand as a possible source of information regarding these evolutionary processes.

One of the most important implications of the above conclusion relates to the use of $\delta^{34}S$ distributions as a function of geological time in identifying the onset of biogenic sulfate reduction (e.g., (13)). Although, in such studies, deposits rich in base-metal sulfides are often excluded because of the potentially complicating effects of hydrothermal processes (surprisingly, the Cu-rich Isua sediments generally are not excluded), highly pyritic but base-metal poor deposits are frequently included. If our deductions from the Recent are correct, then these pyrite-rich rocks may also have formed under conditions where hydrothermal processes were more important than biogenic processes. Thus, the sulfur-isotopic ratios of these rocks may also tell us little about biogenic processes. Sediments which contain sulfide concentrations comparable to those found in normal modern environments are likely to be useful. Typically, these contain no more than about 5 wt.% pyrite.

The study of the isotopic composition of these relatively sulfide-poor rocks may show that sulfate respirers on Earth are older than inferred to date (about 2.75 Gyr, (13)) on the basis of studies of sulfide-rich sediments.

OTHER QUESTIONS OF MORE GENERAL INTEREST

Other questions of a more general, but nevertheless important, nature were raised concerning various aspects of modern and ancient stratified sulfide deposits and microbiological processes. The more important and interesting of these that remain to be answered include:

1) How extensively did the sulfuretum processes which we recognize today function during Precambrian times? Were the processes more or less efficient in the past than they are today?
2) Why do we find $\delta^{34}S$ fractionations up to about 35‰ in laboratory studies involving growing bacteria which perform dissimilatory sulfate reduction, whereas in nature we find the same reduction giving fractionations of up to about 60‰?
3) What is the importance of thiosulfate and of other sulfur compounds of intermediate oxidation state in both Recent and ancient sulfureta?
4) Are there any methods by which we can identify the sources and nature of organic matter in ancient stratified sulfide deposits?
5) How can we best go about defining the evolution of bacteria involved in the sulfur cycle? When did photosynthetic organisms and sulfate reducers first appear?
6) How important is anaerobic methane oxidation now and in the Precambrian?
7) When did abundant sulfate first appear in the Earth's hydrosphere?
8) What criteria might be useful in differentiating sulfides in sediments that formed syngenetically, diagenetically, or epigenetically?

9) Is there any connection between the disappearance of Lake Superior-type banded iron formations at about 2.0 Gyr and the appearance of the shale-hosted stratiform Pb-Zn deposits at about the same time? If there is, what is the nature of the connection?

10) What factors are important in controlling the availability of metals to the environments of formation of stratified sulfide deposits? How has this availability varied through geological time?

REFERENCES

(1) Berner, R.A. 1972. Sulfate reduction, pyrite formation and the oceanic sulfur budget. In The Changing Chemistry of the Oceans (Nobel Symposium 20), eds. D. Dyrssen and D. Jagner, pp. 347-361. Stockholm: Almquist and Wiksell.

(2) Chambers, L.A., and Trudinger, P.A. 1979. Microbiological fractionation of stable sulfur isotopes: a critical review. Geomicrobiol. J. 1: 249-293.

(3) Degens, E.T., and Ross, D.A. 1969. Hot Brines and Recent Heavy Metal Deposits in the Red Sea. New York: Springer-Verlag.

(4) Goldhaber, M.B., and Kaplan, I.R. 1974. The sulfur cycle. In The Sea, ed. E.D. Goldberg, vol. 5, pp. 569-655. New York: John Wiley and Sons.

(5) Hallberg, R. 1980. Diagenetic and environmental effects upon heavy metal distribution in sediments - a hypothesis, with an illustration from the Baltic Sea. In The Dynamic Environment of the Ocean Floor, eds. K.A. Fanning and F.T. Manheim. Lexington: D.C. Heath and Co.

(6) Hekinian, R. et al. 1980. Sulfide deposits from the East Pacific Rise near 21°N. Science 207: 1433-1444.

(7) Jørgensen, B.B. 1980. Mineralization and the bacterial cycling of carbon, nitrogen, and sulphur in marine sediments. In Contemporary Microbial Ecology, eds. D.C. Ellwood, J.N. Hedger, M.J. Latham, J.M. Lynch, and J.H. Slater, pp. 239-251. London: Academic Press.

(8) Krouse, H.R., and McCready, R.G.L. 1979. Reductive reactions in the sulfur cycle. In Biogeochemical cycling of mineral-forming elements, eds. P.A. Trudinger and D.J. Swaine, pp. 315-368. Amsterdam: Elsevier.

(9) Lonsdale, P.F.; Bischoff, J.L.; Burns, V.M.; Kastner, M.; and Sweeney, R.E. 1980: A high-temperature hydrothermal deposit on the seabed at a Gulf of California spreading center. Earth Planet. Sci. Lett. 49: 8-20.

(10) Ohmoto, H., and Rye, R.O. 1979. Isotopes of sulfur and carbon. In Geochemistry of Hydrothermal Ore Deposits, second edition, ed. H.L. Barnes, Ch. 10, pp. 509-567. New York: John Wiley and Sons.

(11) Rickard, D.T. 1973. Limiting conditions for synsedimentary sulfide ore formation. Econ. Geol. 68: 605-617.

(12) Sangster, D.F. 1976. Sulphur and lead isotopes in stratabound deposits. In Handbook of Stratabound and Stratiform Ore Deposits, ed. K.H. Wolf, vol. 2, pp. 219-266. Amsterdam: Elsevier.

(13) Schidlowski, M. 1979. Antiquity and evolutionary status of bacterial sulfate reduction: sulfur isotope evidence. Origins of Life 9: 299-311.

(14) Shanks, W.C., and Bischoff, J.L. 1980. Geochemistry, sulfur isotope composition, and accumulation rates of Red Sea geothermal deposits. Econ. Geol. 75: 445-459.

(15) Skinner, B.J. 1979. The many origins of hydrothermal mineral deposits. In Geochemistry of Hydrothermal Ore Deposits, second edition, ed. H.L. Barnes, Ch. 1, pp. 1-21. New York: John Wiley and Sons.

(16) Spiess, F.N. et al. (RISE Project Group) 1980. East Pacific Rise: Hot springs and geophysical experiments. Science 207: 1421-1432.

(17) Sweeney, R.E., and Kaplan, I.R. 1973. Pyrite framboid formation: laboratory synthesis and marine sediments. Econ. Geol. 68: 618-634.

(18) Trudinger, P.A. 1976. Microbiological processes in relation to ore genesis. In Handbook of Stratabound and Stratiform Ore Deposits, ed. K.H. Wolf, vol. 2, pp. 135-190. Amsterdam: Elsevier.

(19) Trudinger, P.A.; Lambert, I.B.; and Skyring, G.W. 1972. Biogenic sulfide ores; a feasibility study. Econ. Geol. 67: 1114-1127.

Group on Reduced Carbon Compounds in Sediments

Standing, left to right: Jürgen Hahn, Dave McKirdy, Manfred Schidlowski, John Schopf, John Oehler, Steve Golubic, Jochen Hoefs, Alfred Hollerbach, Wolfgang Krumbein.
Seated: Jan-Dirk Arneth, Geoff Eglinton, Christian Junge, John Hayes.

Mineral Deposits and the Evolution of the Biosphere, eds. H.D. Holland and M. Schidlowski, pp. 289-306. Dahlem Konferenzen, 1982.
Berlin, Heidelberg, New York: Springer-Verlag.

Reduced Carbon Compounds in Sediments
State of the Art Report

J. H. Oehler, Rapporteur
J.-D. Arneth, G. Eglinton, S. Golubic, J. H. Hahn,
J. M. Hayes, J. W. Hoefs, A. Hollerbach, C. E. Junge,
W. E. Krumbein, D. M. McKirdy, M. Schidlowski, J. W. Schopf

INTRODUCTION

As suggested by the title of this volume, there is a widespread suspicion that certain types of Precambrian sedimentary mineral deposits may in part owe their origin to the activity of contemporaneous microorganisms. If this view is indeed correct, it is of both economic and academic importance.

The most direct source of information for resolving this issue is the organic matter preserved in rocks of Precambrian age. Organic geochemical and micropaleontological analyses of this organic matter provide a data base, and modern principles of evolutionary theory and microbial ecology provide the conceptual basis for interpreting the data. Thus the sedimentary organic matter tells us something about the kinds of microorganisms that were present in the past, and an understanding of evolutionary processes and microbial ecosystems provides a framework for inferring how such organisms may have interacted with one another and with their environment.

The specific question of whether microorganisms may have contributed to ore-forming processes must be seen as part of the larger topic of the nature and course of biological evolution during the Precambrian. Accordingly, a substantial portion of our discussions concerned the current knowledge of this larger topic.

During the past fifteen years or so, a considerable body of data on the organic geochemistry and micropaleontology of Precambrian sediments has accumulated. It is not the purpose of this paper to review these data or the interpretations drawn from them (for such reviews, see the papers by Awramik, Schidlowski, and McKirdy and Hahn in this volume). Instead, this paper deals with some of the major problems involved in gathering and interpreting the data and with some possible solutions to those problems.

DATA FROM KEROGEN

Kerogen is that portion of the organic matter in rocks which is insoluble in organic solvents. At least 90 percent of the total organic matter (TOM) in most rocks is kerogen. It can be broadly divided into two categories: kerogen which is structurally preserved as recognizable fossils (mostly microfossils in the Precambrian) and kerogen which occurs as particulate debris of a generally amorphous appearance.

Several methods are available for analyzing kerogen. These include: transmitted, fluorescent, and reflected light microscopy; scanning and transmission electron microscopy; elemental analysis of C, H, N, O, S, and P content, among others; isotopic composition of C, H, N, O, and S; and various degradational techniques, such as ozonolysis, pyrolysis, and pyrolysis combined with gas chromatography (GC) or with GC plus mass spectrometry (MS).

We can illustrate some of the problems involved in interpreting data on kerogen by using the example of information obtained from carbon isotopic analyses.

Results of carbon isotopic studies, in conjunction with micropaleontological analyses of the same rocks (see Awramik, this volume), indicate that living organisms existed at least as early as 3.5 Ga ago (Ga = 10^9 years). There is graphitic kerogen in rocks as old as 3.8 Ga from Isua (West Greenland), but because these rocks have experienced numerous episodes of metamorphism, there was a strong opinion in our group that the carbon isotopic data can neither confirm nor deny the presence of life at that time. However, some group members felt that the evidence presently available on both reduced and carbonate carbon in the Isua meta-sediments favors a biological pedigree of these oldest organics and, accordingly, a biological control of the carbon cycle from 3.8 Ga ago (cf. (12)). The graphitic kerogen itself is insufficient evidence for the presence of life, because the organic precursors of the kerogen could have come from nonbiological sources, such as abiotically synthesized organic compounds (see Miller, this volume), carbonaceous chondrites (the flux of which was probably much higher during the early Precambrian than today), or metamorphic alteration of carbonate minerals.

The carbon isotopic data from the 3.5 Ga old rocks suggest that the precursor organisms were autotrophic, but it is unclear what metabolic pathways for carbon incorporation they used. In particular and of special importance to the subject of this conference, it is unclear whether any of these metabolic pathways produced free oxygen.

Between about 3.4 and 2.9 Ga ago, the known sedimentary record is sparse, and the isotopic data are correspondingly limited.

The preserved sedimentary record becomes much more abundant at about 2.9 Ga ago, and the isotopic data for rocks of this age span an exceptionally wide range of values (about -10 to -50‰ PDB). It is tempting to interpret this wide range of values as reflecting microbial "experimentation" with a large number of metabolic pathways using a variety of resource pools. The steadily narrowing range of values from 2.9 Ga to about 2.0 Ga (and, possibly, up to the Phanerozoic) could then be interpreted as representing a gradual abandonment of less advantageous pathways and a "natural selection" toward the pathways characteristic of Phanerozoic organisms. Alternatively, the wide range of isotopic values at about 2,9 Ga ago could be associated with a major increase in the diversity of environments (e.g., the advent of aerobic environments or of cratonic shelf environments) and the ways in which organisms dealt with these new ecological settings. The narrowing range of values toward the Phanerozoic might then be interpreted as reflecting an approach to geological, ecological, and biological "equilibrium," which culminated near the beginning of the Phanerozoic.

In reality, however, the true significance of the isotopic data is not clear. The reason it is unclear is that not enough ancillary data are available to assess the extent to which variables other than metabolic pathways or environmental diversity may have affected the isotopic composition of the organic matter. These variables, which affect not only the isotopic composition but also most of the other chemical characteristics of sedimentary organic matter, include the following:

1) The kind of microbial populations from which the organic matter was derived. This is the primary control over the chemistry of the preserved residue.

2) The depositional environment and lithology of the host rock. These can influence both the original nature of the organic matter and the course of its diagenetic alteration.

3) The thermal maturity of the organic matter.

4) The porosity and permeability of the host rock and the post-depositional environments to which it has been exposed. These influence the degree to which the organic matter may be contaminated by younger organic compounds or altered by low-temperature processes, such as exposure to meteoric waters. They are particularly important factors when dealing with outcrop samples.

All of these variables can affect the type and condition of the organic matter in sedimentary rocks. If we want to discern evolutionary trends and biological processes through analysis of preserved organic matter, we should collect suites of samples in which these variables are controlled or held constant. Barring that, we must at least try to determine the extent to which each variable has affected the organic matter. A variety of techniques and very careful selection of samples are required. One type of geochemical data by itself is insufficient to permit a confident interpretation.

Our inability to interpret confidently the potentially very exciting isotopic data noted above is due to the fact that the effects of some of the variables listed here could not be assessed.

DATA FROM EXTRACTABLE ORGANIC MATTER

Extractable organic matter (EOM) is that portion of the organic matter in a rock that can be removed by treatment with normal organic solvents (i.e., that portion which is not kerogen). It constitutes a small percentage of the total organic matter in most rocks, usually less than 10 percent. However, because it can be brought into solution, it is amenable to analysis by a very broad range of chemical techniques.

The single greatest problem in interpreting data from the EOM in rocks is the difficulty of distinguishing the EOM that was deposited with the original sediment (or was generated within the rock from original organic matter) from EOM which may have been introduced later.

It is widely believed that syndepositional (or syngenetic) EOM contains valuable chemical information regarding ancient organisms and their environments. However, to obtain this information, we must be able to differentiate syngenetic EOM from syngenetic EOM contaminated with postdepositional (or epigenetic) EOM.

Several attempts have been made to overcome this problem of contamination. These have included the progressive extraction of organic matter from rocks ground to increasingly finer particle sizes, the use of rocks that contain no "obvious" contaminants, such as fatty acids and sugars, and the use of unmetamorphosed rocks that contain a relatively high concentration of EOM, i.e., more than 100 ppm. None of these precautions has achieved universal acceptance for consistent reliability.

At this meeting it was suggested that improvements could be achieved by the application of modern methods of petroleum source rock analysis, together with comparative analyses of varying lithologies within a single stratigraphic sequence, and the use of core samples rather than outcrop samples. These methods are explained below.

The petroleum industry's methods of evaluating potential source rocks include guidelines for recognizing the presence of migrated (i.e., postdepositional) organic compounds. These guidelines are based on knowledge of the quantity and the composition of EOM that can be generated from the several types of kerogen during the various stages of maturation (see Tissot and Welte (14)). Abnormally high concentrations of EOM and/or compound distributions that are inconsistent with the type and maturity of the contained kerogen are signs of contamination. Some of the standard methods may require modification if they are applied to metamorphic rocks where postdepositional EOM has been converted to insoluble pyrobitumen (a type of kerogen).

The rationale for analyzing the EOM in a series of rocks of differing lithofacies but from the same stratigraphic sequence is the following. Fine-grained, kerogen-rich rocks (such as shales) generate and retain more EOM than do coarse-grained, kerogen-poor rocks (such as sandstones). We should therefore find more EOM in the fine-grained than in the coarse-grained rocks of an uncontaminated sequence of strata. However, postdepositional (epigenetic) organic compounds should enter more easily the relatively permeable, coarse-grained rocks than the relatively impermeable, fine-grained rocks. Consequently, a contaminated sequence of strata should have quite a different distribution of EOM concentrations than an uncontaminated sequence. Studies based on this kind of relationship, however, must consider the possibility that sandstones can contain EOM derived from an adjacent shale (the source-reservoir principle of petroleum geochemistry). Strictly speaking, such a sandstone is "contaminated," but in our nomenclature, the contaminant is the syngenetic EOM of the shale.

The advantage of using core samples rather than outcrop samples is that deeply buried rocks (and especially those that have never been at the surface) are less likely to be contaminated than those that are exposed at the surface. Unfortunately, core samples of rocks that are of greatest interest are not always available. Quarry samples, or samples taken by means of hand-held coring machines or inorganic explosives, are obvious alternatives.

A combination of these methods has apparently been used only once (7); the results are encouraging and interpretations based on the organic geochemical and organic petrographical data seem thoroughly consistent with the geological and mineralogical data.

If we can circumvent the contamination problem in this way, then we can gain new information from sedimentary organic compounds, and particularly from biological marker compounds and from their diagenetic and catagenetic products. Among these, murein which is associated with eubacteria, chlorophyll associated with oxygen-producing photosynthesizers, bacteriochlorophyll associated with photobacteria, and possibly C_{21+} acyclic isoprenoid alkanes associated with archaebacteria are particularly important. These compounds, with the exception of the alkanes, could not have survived unaltered in sedimentary rocks since Precambrian times, but the recognition of their diagenetic and catagenetic products as syngenetic constituents of Precambrian organic matter would go far toward establishing the presence of their original source-organisms at the time of sedimentation.

To assist in establishing and validating the Precambrian organic geochemical record, we must explore the full Phanerozoic record, from modern depositional environments backwards to the Precambrian to establish maturation decay curves for extractable organic matter (EOM) and kerogen. In this way we are apt to recognize most clearly the detailed molecular information remaining in ancient sediments that bears on the reconstruction of ancient environments.

DATA FROM SELECTED MODERN MICROBIAL ECOSYSTEMS

The enormous time elapsed since the Precambrian and the shift in the nature of the biota limit the use of recent-to-fossil comparisons. Nevertheless, microbial ecosystems similar to those found in the Precambrian fossil record continue to function in particular modern environments with little interference from organisms of higher organizational complexity and later evolutionary origin.

Three papers in this volume review the current state of knowledge concerning some of the modern microbial ecosystems that appear to have important analogues in the Precambrian (those by Edmunds et al., Nealson, and Trüper). Only a few important points will be highlighted here.

Although photoautotrophs are the primary producers of organic matter in most laminated microbial ecosystems (e.g., stromatolites), heterotrophic degraders comprise a significant proportion of the organisms present in these ecosystems. Commonly the degraders, which are mostly bacteria, are so small that their remnants, even if they were structurally preserved, might not be recognized in Precambrian sediments. Much morphological and chemical destruction of cells occurs prior to burial and lithification of microbial mats. Stratification of populations of vastly varying physiologies (e.g., oxygen producers, sulfate reducers, methanogens, and so on) occurs over very short vertical (and horizontal) distances. In the process of sediment accretion, these physiological zones move as progressive "degradational fronts" leaving little of the original organic input behind. In modern microbial mats, it is this residue which contains biochemical and isotopic information comparable to that potentially preserved in fossil organic matter. In addition, fossil organic matter has, of course, been altered by diagenesis and maturation.

Each of these points has significant implications for our interpretation of the Precambrian fossil record and for inferences concerning possible relationships between biological activity and the formation of Precambrian mineral deposits.

PRECAMBRIAN MICROORGANISMS AND ORE-FORMING PROCESSES

Despite the limitations emphasized above, we do feel that data currently available permit several conclusions concerning biological involvement in the formation of certain Precambrian mineral deposits. Our main conclusion is that we see no compelling evidence for a causal relationship between the activity of microorganisms and the deposition of banded iron formations (BIF's), phosphorites, or stratiform sulfide deposits. To be sure, there is evidence of contemporaneous biological activity in the basins where some of these deposits were forming (e.g., (10)). This, however, only means that the environments in the basins were congenial to life; it does not necessarily mean that organisms were responsible for mineral deposition or that the mineral deposits would not have formed without the presence of organisms.

Perhaps the most difficult deposits to evaluate in this regard are the giant BIF's of the Hamersley/Cape Province/Transvaal type. It has long been thought by several workers in the field that substantial (though fluctuating) quantities of dissolved oxygen in the water column were required for the precipitation of ferric oxides in these deposits, and that such quantitites of oxygen could only have been generated by oxygen-producing, photosynthetic organisms. It is now unclear whether biologically produced oxygen was indeed required (see Button et al., this volume) for ferric oxide precipitation.

Nevertheless, the time of the first oxygen-producing photosynthesis, of the first aerobic environments, and of the first oxidizing atmosphere remain outstanding problems in evolutionary biology regardless of their relevance to the formation of BIF's. These questions were addressed, and some means of resolving them were proposed.

It is likely that the total standing biomass in an anoxic world (such as that which seems to have existed during the early Precambrian) would have been substantially less than that existing in the presence of atmospheric oxygen. It is uncertain, however, whether the transition from anoxic to oxic conditions produced a "quantum jump" in the quantity of organic matter preserved in the sedimentary record since the fraction of organic matter preserved should have been higher when the earth was anoxic. Evidence for the transition from anoxic to oxidizing conditions might be recorded more convincingly in the organic content of ancient sandstones than in that of shales. Under oxidizing conditions, sandstones contain little organic matter; under anoxic conditions, their content of organic matter might have been more equal to that of shales.

The biological nitrogen cycle during anoxic times may have been different from that which evolved in response to the advent of oxidizing conditions; this difference might be reflected in the degree of biological nitrogen isotope fractionation. Analyses of the isotopic composition of nitrogen in preserved organic matter might therefore indicate the time of origin of aerobic, aqueous environments.

Certain nitrogen-fixing algae possess enlarged cells (heterocysts), which protect the nitrogen-fixing apparatus from oxygen. If bona fide heterocysts were identified in the fossil record, they would strongly suggest the presence of free oxygen in the water column during the lifetimes of those particular algae.

Finally, there are a number of potential biological marker compounds (such as porphyrins and hopenoids) that could be used to distinguish microorganisms that produce oxygen from those that do not. If identified in the syngenetic EOM of Precambrian sediments, these compounds could document the presence or absence of oxygen-producing photosynthesizers.

NEEDS

From the foregoing, it is evident that our knowledge of Precambrian sedimentary organic matter is far from perfect. To help improve that situation, we have defined four fundamental requirements.

1) Improved communication among scientists in the numerous disciplines that contribute to our knowledge of the nature and significance of sedimentary organic matter, including more precisely defined terminology and generally accepted terminological rules for naming and describing new observations;

2) The financial means for increased international, interdisciplinary cooperation and collaboration;

3) Increased interaction with petroleum and mining companies to promote sample exchange and technology transfer; and

4) Carefully collected geological and biological samples including, in particular, geochronologically well-documented, unweathered samples from (a) a range of well-defined depositional environments; (b) individual strata that have experienced varying degrees of metamorphic alteration; and (c) unmetamorphosed stratigraphic sequences older than 3.0 Ga.

Each of these (with the possible exception of 4c) is a requirement that can be fulfilled realistically in the near future if deliberate action is taken.

OUTSTANDING UNSOLVED PROBLEMS

The following list is phrased in the form of questions, each of which represents an area of research that should be pursued.

1) How can biotic organic matter be distinguished from abiotic organic matter in ancient sediments, and what is the temporal distribution of the latter?

2) How can we better distinguish syngenetic EOM from epigenetic EOM?

3) By what methods can the chemical composition of individual microfossils be characterized?

4) What differences are there between organic materials preserved in coeval but differing Precambrian sedimentary (depositional) facies?

5) What is the nature and significance of organic matter preserved in sedimentary ore deposits?

6) What organic geochemical indicators can be used to distinguish aerobic-anaerobic from solely anaerobic ecosystems as preserved in the geologic record?

7) What secular trends and major evolutionary events are recorded in Precambrian sedimentary organic matter?

8) Have we adequately documented the physical extent of the present biosphere, particularly in regard to extreme environments? For example, to what depths in sedimentary columns does microbiological activity persist, what effects does that activity have on sedimentary organic matter, and what imprint does this impose on the organic geochemical record?

9) What are the detailed correlations between elemental and isotopic compositions of organic matter, the physical and structural characteristics of organic matter, the mineralogy of the enclosing sediment, and the depositional histories of those materials, particularly in the Precambrian? For example, what changes occur in the elemental and isotopic composition of organic matter subjected to increasing maturation within different host rock lithologies? Which of these correlations, if any, can be discerned by microscopic techniques, and how?

10) What information on the nature of the Precambrian biosphere, hydrosphere, and atmosphere can be gained from the contents of primary fluid inclusions?

11) What are the organic geochemical characteristics of Precambrian phosphorites?

12) To what extent do C, H, N, O, and S isotopes in organic matter exchange with the environment during diagenesis and maturation?

13) In terms of microbial ecosystems, to what extent is the present the key to the past?

MEASUREMENTS TO BE MADE

To help solve some of the problems enumerated above, we recommend the following procedures and measurements.

1) Microfossils found in macerations should also be studied in thin sections.

2) Organic geochemical studies of isolated kerogens should be accompanied by studies of the same kerogens in thin and polished sections.

3) Kerogens should be analyzed by x-ray diffraction and by fluorescence and reflectance microscopy.

4) In samples in which the isotopic composition of carbon is measured, the isotopic composition of hydrogen and the atomic ratios H/C, O/C, N/C, and P/C should also be measured.

5) The atomic ratios H/C, O/C, N/C, and P/C should be determined in all samples analyzed by pyrolysis-GC and pyrolysis-GC-MS.

6) Efforts should be made to determine the biological and biochemical nature of organisms grown in anoxic atmospheres.

7) The isotopic, elemental, and molecular composition of microbial mats degraded aerobically and anaerobically should be compared with the composition of the same mats degraded only anaerobically.

8) The isotopic and chemical composition of pure cultures of microorganisms should be analyzed and should be searched for the presence of potential biological marker compounds with the aim of identifying the environmentally and phylogenetically significant molecules and structures synthesized by present-day organisms.

9) Artificial maturation studies of microorganisms and microbial organic matter should be conducted; their maturation, including changes in stereochemical and isotopic properties, should be traced and fully documented as a function of time, temperature, and mineral matrix. All possible methods should be used to get this information out of cellular material on its parallel and interwoven paths through organic debris, protokerogen, kerogen, and extractable organic matter.

10) Maturation studies of individual compounds such as pristane and phytane should be conducted in which changes in stereochemistry are correlated with changes in carbon and hydrogen isotopic composition.

11) Detailed survey measurements of biological marker compounds and assemblages of compounds characteristic of phylogenetically significant organisms (particularly microorganisms) should be greatly extended. The aims should be to establish inventories of primary biological markers and the progressive diagenetic products of those markers so that fossil molecules can be better correlated to original biological sources.

12) Carbon isotopic measurements of organic compounds should be made at both the inter- and intramolecular levels.

13) Attempts should be made to analyze the contents of individual, primary fluid inclusions.

Several important tasks consequent upon the above can be defined:

1) selection of phylogenetically key organisms (those which are characteristic of particular environments and which are thought to have long fossil records) for detailed molecular analyses, and construction of a data base (preferably computerized) for the compounds identified in relation to the organisms examined and their phylogeny, habitats, and environmental significance;

2) a parallel documentary study of selected, restricted modern environments and of past equivalents which are best described in terms of their sedimentology, biological inputs, temperature histories, oxicity/anoxicity, and other physical and chemical parameters; and

3) documentation and quantification of maturation parameters based on the molecular information obtained from 1) and 2).

Thus, we may gain answers to the questions of which suites of molecular markers carry their message furthest back through the geologic record and in which rock types (mineral matrices) they are best preserved.

TECHNIQUES AND TOOLS

The following techniqes and instruments should be used to achieve the measurements outlined above:

1) Ion microprobe analysis of isolated and especially in situ microfossils so that organic chemistry can be correlated with fossil morphology.

2) Separation of large quantities of monospecific microfossil assemblages and individual kerogen types so that "pure" rather than mixed samples of material can be analzyed as an adjunct to technique No. 1 above.

3) Stepwise pyrolysis-GC and pyrolysis-GC-MS analysis of kerogens.

4) High resolution mass spectrometry (HRMS) of organic compounds.

5) Determination of the intramolecular distributions of ^{12}C and ^{13}C in organic compounds.

6) Photo-acoustic spectroscopy.

7) Laser pyrolysis and laser ionization mass spectroscopy of in situ microfossils and individual kerogen particles.

8) Use of microelectrodes to study microbial environments.

9) Identification and use of additional specific reagents for degrading kerogens.

10) Use of standardized laboratory microbial ecosystems for experimental studies.

11) Utilization, so far as possible, of methods that permit digital data recording, such as photometric microscopy.

12) Use of statistical data analysis whenever possible.

13) Use of high performance liquid chromatography and HPLC-mass spectrometry to analyze alkanes and more polar compounds.

14) Separation of individual compounds (using high resolution capillary-GC) in quantities sufficient for precise structural determinations by techniques such as ^{13}C nuclear magnetic resonance, high resolution mass spectrometry, and X-ray diffraction.

15) Field measurements of nitrogen fixation in microbial ecosystems.

16) Judicious use of image-enhancement techniques.

GENERAL CONCLUSIONS

A considerable body of information has been accumulated in the fields of Precambrian micropaleontology, organic geochemistry, and microbial ecology. This information has allowed us to draw a number of conclusions concerning the nature and significance of organic matter in Precambrian sediments, and most of these conclusions are to be found in the papers of this volume. At this meeting we concluded that there is no compelling evidence for a causal relationship between Precambrian biological activity and the formation of economic mineral deposits. Our discussions demonstrated how little we know about Precambrian life and how much remains to be learned. In bringing together so many people of such diverse disciplinary backgrounds, this conference has helped us to define what additional information can be gained, how we can go about getting it, and what kinds of collaborative research programs might best be undertaken to obtain the massive amount of data required to permit confident interpretations of the early evolutionary history of life and the nature of biological interactions with Precambrian environments.

REFERENCES

The following is a list of general references, not necessarily cited here, to which the interested reader can refer for a broader introduction to the subjects discussed.

(1) Brock, T.D. 1979. The Biology of Microorganisms. Englewood Cliffs, NJ: Prentice Hall.

(2) Golubic, S. 1976. Organisms that build stromatolites. In Stromatolites: Developments in Sedimentology, ed. M.R. Walter, vol. 20, pp. 113-126. Amsterdam: Elsevier.

(3) Golubic, S. 1980. Early photosynthetic microorganisms and environmental evolution. In GOSPAR Life Sciences and Space Research, ed. R. Holmquist, vol. 5, pp. 101-107. London: Pergamon.

(4) Holzer, G.; Oro, J.; and Tornabene, T.G. 1979. Gas chromatographic/mass spectrometric analysis of natural lipids from methanogenic and thermoacidophilic bacteria. J. Chromatog. 186: 873-877.

(5) Krumbein, W.E., ed. Microbial Geochemistry. Blackwell, in press.

(6) McKirdy, D.M. 1974. Organic geochemistry in Precambrian research. Precambrian Res. 1: 75-137.

(7) McKirdy, D.M., and Kantsler, A.J. 1980. Oil geochemistry and potential source rocks of the Officer Basin, South Australia. Austral. Petrol. Explor. Asso. J. 20: 68-86.

(8) Oehler, D.Z., and Smith, J.W. 1977. Isotopic composition of reduced and oxidized carbon in early Archaean rocks from Isua, Greenland. Precambrian Res. 5: 221-228.

(9) Oehler, J.H. 1976. Experimental studies in Precambrian paleontology: structural and chemical changes in blue-green algae during stimulated fossilization in synthetic chert. Geol. Soc. Am. Bull. 87: 117-129.

(10) Oehler, J.H., and Logan, R.G. 1977. Microfossils, cherts, and associated mineralization in the Proterozoic McArthur (H.Y.C.) lead-zinc-silver deposit. Econ. Geol. 72: 1393-1409.

(11) Ourisson, G.; Albrecht, P.; and Rohmer, M. 1979. The hopenoids: Paleochemistry and biochemistry of a group of natural products. Pure and Applied Chem. 51: 709-729.

(12) Schidlowski, M.; Appel, P.U.V.; Eichmann, R.; and Junge, C.E. 1979. Carbon isotope geochemistry of the 3.7 x 10^9 yr old Isua sediments, west Greenland: implications for the Archaean carbon and oxygen cycles. Geochim. Cosmochim. Acta 43: 189-199.

(13) Schopf, J.W. 1975. Precambrian paleobiology: problems and perspectives. Ann. Rev. Earth Planet. Sci. 3: 213-249.

(14) Tissot, B., and Welte, D.H. 1978. Petroleum Formation and Occurrence. Heidelberg: Springer-Verlag.

Group on
Biogeochemical Evolution of the Ocean-Atmosphere System

Standing, left to right: Abe Lerman, Hugh Jenkyns, Jan Veizer, Charles Curtis, John Langridge, Arie Nissenbaum, Stan Miller. Seated: Bob Folinsbee, Stan Awramik, Dick Holland, Preston Cloud.

Mineral Deposits and the Evolution of the Biosphere, eds. H.D. Holland and M. Schidlowski, pp. 309-320. Dahlem Konferenzen, 1982.
Berlin, Heidelberg, New York: Springer-Verlag.

Biogeochemical Evolution of the Ocean-Atmosphere System State of the Art Report

S. M. Awramik, Rapporteur
P. Cloud, C. D. Curtis, R. E. Folinsbee,
H. D. Holland, H. C. Jenkyns, J. Langridge,
A. Lerman, S. L. Miller, A. Nissenbaum, J. Veizer

INTRODUCTION

Scope of the Report

The present composition of the atmosphere-ocean system is the result of the extremely complex interaction of crustal and biological evolution. Oceanic chemistry and its evolution reflect the contribution of: a) mantle-derived material where new oceanic crust is formed, b) the input of continentally-derived material, c) the impact of biological systems, and d) the interactions with the atmosphere.

It can be argued that the appearance of oxygen-producing photosynthetic organisms was the most important event in the evolution of the atmosphere-ocean system. It is mainly for this reason that our group concentrated on the evolution of atmospheric oxygen. The event must have occurred early in Earth history; we do not know precisely when, but the production of molecular oxygen has consequences that can be tracked in the geological record.

ATMOSPHERIC OXYGEN - PRESENT AND PAST

The problem is not to explain the absence of atmospheric oxygen on the early Earth but to explain how, when, and why oxygen took up permanent residence in the atmosphere.

Two extremely different atmospheric oxygen regimes were initially considered: 1) an oxygenic atmosphere such as that of the present day composed of ~20% O_2 and 2) an anoxic atmosphere such as was almost certainly required for the origin of life some 4000 Ma ago (1 Ma = 10^6 years). We agreed that oxygen-producing photosynthesis was and is the major source for atmospheric oxygen; photolysis of water vapor in the upper atmosphere on the primitive Earth must have produced some molecular oxygen, but there is considerable debate concerning the importance of this mechanism (21). In the absence of an ozone screen, the production of O_2 by photolysis would not have been limited to the upper atmosphere, but the appearance and maintenance of oxygen in the atmosphere depended on the relative magnitude of the rate of O_2 production and the rate of O_2 consumption by a variety of mechanisms.

1.2×10^{21} g O_2 are present in the atmosphere today. The rate at which oxygen is produced as a product of the burial of organic matter (~4×10^{14} g O_2/yr) is almost equal to the rate at which oxygen is consumed by oxidation processes (~4×10^{14} g O_2/yr) (13). Blue-green algae (cyanobacteria), algae, and plants are the oxygen producers. Oxygen consumption occurs during the oxidation of volcanic gases (~1×10^{14} g O_2/yr consumed) and by rock weathering (~3×10^{14} g O_2/yr consumed), principally by the oxidation of reduced carbon, sulfide, and ferrous iron.

At the other end of the time scale, we chose the period of the prebiotic synthesis of organic compounds and the origin of life, more than 3800 Ma ago. It is generally accepted that at this time there was no free O_2 in the atmosphere ($<10^{-10}$ atm), because reducing conditions are required for the synthesis of the organic compounds needed for the development of life, and because an analysis of the H_2 balance suggests that free hydrogen was present in the atmosphere at pressures on the order of 10^{-3} atm.

The time of the origin of life is uncertain. It must have occurred before the earliest definite evidence of life for the existence of living organisms on Earth (3500 Ma ago), but may have taken place much earlier. One need not assume that long periods of time were required for the synthesis of life. A period of 10^7 to 10^8 years or less following the time after the Earth had cooled sufficiently for the requisite organic compounds to be stable may well have been adequate. The temperature at which the surface of the Earth had "cooled down sufficiently" is difficult to estimate, but is probably less than 100°C. If higher surface temperatures persisted until 4000 Ma ago, then life probably originated about 3900 Ma ago.

The necessity for the synthesis of organic compounds on the primitive Earth places a number of constraints on the chemistry of the early atmosphere. The behavior of various gas mixtures when subjected to electric discharges has been studied intensively. The nature of the organic compounds formed in such discharge experiments depends on the composition of the original gas mixtures.

1. CH_4 (methane), NH_3 (ammonia), H_2O (water) with or without H_2 (hydrogen). From this strongly reducing mixture of gases, a wide variety of organic compounds, particularly amino acids, are produced in excellent yield.
2. CH_4, N_2 (nitrogen), H_2O (with a trace of NH_3). This mixture is also quite reducing and yields as wide a variety of organic compounds as mixture #1, including about 13 of the 20 amino acids that occur in proteins. Such an atmosphere is considered more plausible than mixture #1, since NH_3 tends to dissolve in the comtemporary oceans, and/or to be destroyed by photochemical reactions.
3. CO (carbon monoxide), H_2, N_2 (with a trace of NH_3). Such mixtures are less reducing than 1 or 2; they yield amino acids in fair quantities, but the product is largely glycine.
4. CO_2 (carbon dioxide), H_2, N_2 (with a trace of NH_3). The results are similar to 3.
5. CO, H_2O, N_2. No significant yield of organic compounds is obtained in the absence of H_2.

6. CO_2, H_2O, N_2. Same result as 5.

The experimental results obtained so far show that the most efficient organic syntheses occur in the more reducing gas mixtures, that is, in those containing CH_4. CH_4 pressures need only be on the order of 10^{-5} to 10^{-2} atmospheres.

Hence, if the most efficient syntheses of organic compounds is postulated, then a strongly reducing atmosphere is indicated. However, it is not clear that high efficiency was really a requirement for the origin of life. Atmospheres containing CO_2 and CO are favored by some researchers, since these are the major carbon compounds emitted by volcanoes (25). However, volcanoes may have been the only major source of gases for the primitive atmosphere, and the composition of volcanic gases may not have been the sole determinant of the composition of the primitive atmosphere.

Most of the NH_3 in a primordial atmosphere would have dissolved in the contemporary ocean and would have been converted to NH_4^+ if the pH of the ocean was about 8. Remaining atmospheric ammonia would have been rapidly decomposed by solar radiation. The NH_4^+ concentration probably did not rise much above 0.01M because NH_4^+ is strongly absorbed by clay minerals. Its concentration was probably not much lower than 0.01M, because some amino acids are not stable at lower concentrations. An ammonium ion concentration of about 0.01M is needed for many of the polymerization processes believed to have occurred in the conversion of the monomers to polymers on the primitive Earth (3). The source of NH_3 necessary to maintain such concentrations has not been identified. It has been suggested that NH_3 was introduced by meteorites and that it was generated during the decomposition of introduced extraterrestrial amino acids or perhaps by the photochemical reduction of atmospheric nitrogen by a naturally occurring form of TiO_2 (12). Hydrothermal processes could also have been important.

TRACKING PRECAMBRIAN OXYGEN LEVELS - THE GEOLOGICAL RECORD

The ultimate source of information on levels of oxygen in the

Precambrian is data from the geological record. We recognize the following benchmarks:

1. The Isua supracrustals (West Greenland) with an age of ~3800 Ma (the banded ironstone (15)) provide some constraints on the early Archean atmosphere-ocean system. The nature of the metasedimentary rocks indicates that they were deposited in an aqueous environment. Also the sediments were apparently deposited under anaerobic conditions. Carbonate rocks in the metasedimentary sequence (1) imply the presence of atmospheric CO_2. Disagreement exists whether sizeable quantities of CO_2 and CH_4 could coexist in such an atmosphere. If not, then the evidence for the presence of CO_2 indicates that the atmosphere was not strongly reducing. Elemental carbon is present in the Isua supracrustals, but it is not clear whether the carbon is detrital in origin or whether it is the product of biological activity during Isua times. The isotopic composition of the carbon has been used to suggest that it is the product of autotrophic CO_2 fixation (19), but the same isotopic signature could probably have been generated during prebiotic syntheses or by the degree of metamorphism.
2. Filamentous microfossils (2) and stromatolites (14,26) in rocks of the Warrawoona Group of Western Australia are convincing evidence for biological activity 3500 Ma ago. The $\delta^{13}C$ data are consistent with the existence of autotrophic CO_2 fixation (Hayes and others in (2)), but we have no convincing data regarding the metabolism of the contemporary microorganisms. Though some of the microfossils resemble cyanobacteria (or a variety of prokaryotes for that matter), it is not clear whether any of these were actually oxygen-producing photosynthesizers. It is not safe to assume that the physiology of an ancient microorganism was the same as that of its modern morphological analogs.
3. Could O_2 production via photosynthesis in the Archean have been as high as it is today, but that a high rate of reductant supply kept the system nearly anoxic? The concentration of reduced carbon in early Archean sediments of the Swaziland Supergroup is

similar to that in modern sediments, and it has been suggested that primary organic productivity has been roughly constant since the early Archean (16). However, serious questions were raised regarding the validity of this inference. To some, the isotopic composition carbon in Archean and Proterozoic carbonates is consistent with the interpretation that oxygen production was nearly the same then as now and that the oxygen content of the atmosphere may have been close to 80% of the present levels by 3000 Ma ago (18). Assumptions that underlie this conclusion are that the light isotope-enriched reduced carbon is the result of autotrophs, principally oxygen-producing photosynthesizers, and that the ratio of oxidized to reduced carbon (3-4:1) has not changed significantly with time. Unfortunately, even if these assumptions are valid, the available isotopic data can be interpreted equally well in terms of an atmosphere which contained very little free oxygen.

4. Archean and Early Proterozoic banded iron formations have been considered as a major sink for free O_2 (5). We were of the opinion that the major source of this oxygen was oxygen-releasing photosynthesis but, following Button et al. (this volume), one should probably not dismiss the contribution of the photochemical oxidation of ferrous iron and of the photolytic dissociation of water that may have resulted in the oxidation and precipitation of iron. The flux of ferrous iron into Archean ocean basins may have been 2 to 3 times that of the present day (22).

5. Detrital uraninite and pyrite in the 2600 to 2700 Ma-old Witwatersrand (South Africa) and 2200 to 2500 Ma-old Huronian (Canada) sediments are regarded by many to have accumulated under essentially anoxic (see Cloud (7)) or low-P_{O_2} conditions. Detrital uraninite has been found to persist in sediments of the Indus River system (20). However, the special climatic circumstances and high rates of erosion and transport minimize the importance of uraninite survival in this area as evidence against a detrital origin for uraninites in pre-2200 Ma-old rock units.

6. Precambrian weathering profiles are a potential source of information regarding P_{O_2} levels in the contemporary atmosphere. The opinion was expressed by some and doubted by others in the group that under anoxic conditions or low P_{O_2} levels, the rate of chemical weathering is much less than today. 1-2m weathering profile has been found at the base of the Huronian Supergroup (below the detrital uraninite) in Ontario, Canada (8). Could P_{CO_2} levels equal to or greater than the present-day value account for this deep weathering in the absence of oxygenic atmospheric conditions? Land plants today play a major role in chemical weathering by the binding action of their roots. Without this binding action, fine-grained (large surface area, high reactivity) rock debris would be rapidly eroded from slopes in humid climates - where most intensive atmosphere/lithosphere interaction occurs today. The absence of binding agents prior to the Silurian (rooted vascular plants) could thus have been very important. At the same time the oxidation of organic remains in soils today generates CO_2 and speeds chemical weathering. We know very little about organisms which occupied the land surface prior to the Silurian. This uncertainty is important in the context of the oxygen demand during weathering and in the solute flux from the land surface. Research on Precambrian weathering profiles is relatively new, and their potential for providing constraints on P_{O_2} and P_{CO_2} levels in the past remains to be exploited.
7. The Precambrian fossil record provides clues to atmospheric oxygen levels. Cloud (6,7) has summarized the following guidelines which can be used as a barometer of O_2 levels in the Precambrian: a) The appearance of heterocystous filamentous cyanobacteria (the heterocysts are sites of nitrogen fixation) in stromatolitic cherts of the Gunflint Iron Formation (~2000 Ma-old) was a response to oxygenic conditions. However, there is some question whether the observed features are bona fide heterocysts or whether they are heterocyst-like structures produced as a result of the differential shrinkage of a tubular microbe. b) The first appearance of eukaryotes signalled at least 1% PAL (Present Atmospheric

Level) of oxygen, because this quantity of O_2 is needed for mitosis. It is not clear at present just when the first eukaryote appeared. c) Simple metazoans, where O_2 diffuses across membranes require ~7% PAL of oxygen. Ediacaran and related faunas which appeared about 680 Ma ago might have required approximately this level of oxygen. It would be interesting to see whether there are extinctions or adaptive radiations among Precambrian microfossils and macrofossils that might be related to changing levels in atmospheric oxygen.

THE OCEAN SYSTEM

The ocean is a medium which integrates local processes on a global scale. Variations in seawater composition are recorded in the nature of chemical sediments and place constraints on the extent of changes in these processes through time. Postdepositional alteration of these sediments must, of course, be recognized and taken into account (4). Marine evaporites are particularly useful sediments; carbonate rocks are also suited for such studies because they are relatively simple, and because the partition coefficients of some trace elements and isotopes between seawater and carbonates are known. Diagenetic processes in carbonates are understood at least in part (4), background work has been done on their chemistry (22), and they are common in the Proterozoic but, unfortunately, less so in the Archean.

The data obtained from analyzing Sr, Fe, Mn, Ba, Na, Ca, Mg, $^{87}Sr/^{86}/Sr$, $\delta^{18}O$, $\delta^{13}O$ and associated sulfide in late Archean and early Proterozoic carbonate rocks can be interpreted as a relatively high heat flow from the interior and with an ocean chemistry more strongly influenced by mantle fluxes in the Archean than today; in the early Proterozoic, the heat flux decreased, continents grew, and the most important chemical inputs to the oceans shifted more toward continental river discharge (for example, see Veizer (22)).

OUTSTANDING QUESTIONS AND RECOMMENDATIONS

1. What was the effect, if any, of the episode of meteorite bombardment ending about ~3900 Ma ago (9,11) on the atmosphere-ocean system? Could certain organic compounds or trace metals required in the synthesis of life have been introduced and utilized during this episode? Will we be able to find rocks older than 3800 Ma-old that survived the bombardment episode?
2. When did life arise, and how soon after it arose did oxygen-releasing photosynthesis appear? We need some way to distinguish between oxygen-releasing photoautotrophy and other types of autotrophy in the geological record.
3. The Phanerozoic atmospheric P_{O_2} might be controlled in part by the coupled cycles of carbon and sulfur (10,24). This is reflected in the $\delta^{13}C$ value of carbonates and organic carbon and in the $\delta^{34}S$ value of sulfates, sulfides, and perhaps in the changes in the C^o/S ratio in sediments during the Phanerozoic. The carbon and sulfur cycles may have been decoupled more than 2700 Ma ago, and C-Fe coupling may have been more important in controlling atmospheric P_{O_2}. We should perhaps look for complementary isotopic ($\delta^{13}C$) and elemental trends in C, Fe, and Mn in (chemical) sediments. If present, these trends may contain the information necessary to construct reservoir models similar to those devised for the Phanerozoic.
4. What was the availability of phosphorus to life in the Precambrian? What controls the C:P ratio? If PO_4^{3-} input was lower, did it limit the amount of primary productivity? Volcanic ash contains quite a bit of phosphorus (perhaps as much as twice that of other sediments); how might this difference have affected productivity in the Precambrian? The oldest phosphorite deposits known to date are apatite-rich mesobands in 2600 to 2000 Ma-old iron formations (Button et al., this volume).
5. Several models and the absence of Metazoa suggest that the atmosphere was low in oxygen, a few percent PAL, in the interval from about 2000 Ma to the beginning of the Phanerozoic (6). If this is true, why did O_2 remain at such low

levels during this period?

6. The oldest generally recognized red beds, presumably signalling oxygenic atmospheric conditions, are from the ~2300 Ma-old Huronian Supergroup (17). Regarding red beds or red sediments: How do they form? What is the time of their oxidation: before, during, after deposition, or all of these? There are probably examples of each, but how can we tell?
7. Is there any relationship between periods of major volcanism and O_2 levels?
8. We are preoccupied with the construction of models for our understanding of the atmosphere-ocean system and its evolution. We need much more field evidence for our models such as the environments of deposition of banded iron formation, phosphorites, and evaporites. To what extent do these and other deposits reflect global conditions? What are the verifiable consequences of the proposed models?
9. There is little information regarding oxygen requirements for marine organisms. Is it possible to develop new oxygen barometers which can be applied to the fossil record?
10. More data are needed on components in the atmosphere (other than oxygen) that leave geologically identifiable traces. Isotope work on the other gases is needed.

BACKGROUND READING

Cloud, P. 1976. Major features of crustal evolution. Geol. Soc. S. Afr., Annexure 79: 1-32.

Holland, H.D. 1978. The Chemistry of the Atmosphere and Oceans. New York: Wiley and Sons.

Miller, S.C., and Orgel, L.E. 1974. The Origin of Life on Earth. Englewood Cliffs: Prentice-Hall.

Veizer, J. 1976. Evolution of ores and sedimentary affiliation through geologic history; relations to the general tendencies in evolution of the crust, hydrosphere, atmosphere and biosphere. In Handbook of Strata-bound and Stratiform Ore Deposits - III, ed. K.H. Wolf, pp. 1-41. Amsterdam: Elsevier.

Walker, J.C.G. 1977. Evolution of the Atmosphere. New York: MacMillan Publishing.

REFERENCES

(1) Allaart, J.H. 1976. The pre-3760 m.y. old supracrustal rocks of the Isua area, central West Greenland, and associated occurrence quartz-banded ironstone. In The Early History of the Earth, ed. B.F. Windley, pp. 177-189. New York: Wiley and Sons.

(2) Awramik, S.M.; Schopf, J.W.; Walter, M.R.; and Buick, R. Filamentous fossil bacteria 3.5 x 10^9-years-old from the Archean of Western Australia. Science, in press.

(3) Bada, J.L., and Miller, S.L. 1967. Ammonium ion concentration in the primitive atmosphere. Science 159: 423-425.

(4) Brand, U., and Veizer, J. 1980. Chemical diagenesis of a multicomponent carbonate system - 1: Trace elements. J. Sed. Petrol. 50(4).

(5) Cloud, P. 1973. Paleoecological significance of the banded iron formation. Econ. Geol. 68: 1135-1143.

(6) Cloud, P. 1976. Beginnings of biospheric evolution and their biogeochemical consequences. Paleobiology 2: 351-387.

(7) Cloud, P. 1976. Major features of crustal evolution. Geol. Soc. S. Afr., Annexure 79: 1-32.

(8) Frarey, M.J., and Roscoe, S.M. 1970. The Huronian Supergroup north of Lake Huron. In Symposium on Basins and Geosynclines of the Canadian Shield, ed. A.J. Baer, paper 70-40, pp. 143-157. Geological Survey of Canada.

(9) Frey, H. 1980. Crustal evolution of the early Earth: the role of major impacts. Precambrian Res. 10: 195-216.

(10) Garrels, R.M., and Perry, E.A., Jr. 1974. The cycling of carbon, sulfur, and oxygen through geologic time. In The Sea, ed. D. Goldberg, vol. 5. New York: Wiley and Sons.

(11) Goodwin, A.M. 1976. Giant impacting and the development of continental crust. In The Early History at the Earth, ed. B.F. Windley, pp. 77-95. New York: Wiley and Sons.

(12) Henderson-Sellers, A., and Schwartz, A.W. 1960. Chemical evolution and ammonia in the early Earth's atmosphere. Nature 287: 526-528.

(13) Holland, H.D. 1978. The Chemistry of the Atmosphere and Oceans. New York: Wiley and Sons.

(14) Lowe, D.R. 1980. Stromatolites 3,400 m.y. old from the Archean of Western Australia. Nature 284: 441-443.

(15) Myers, J.S. 1976. The early Precambrian gneiss complex of Greenland. In The Early History of the Earth, ed. B.F. Windley, pp. 165-176. New York: Wiley and Sons.

(16) Reimer, T.O.; Barghoorn, E.S.; and Margulis, L. 1979. Primary productivity in an early Archean microbial ecosystem. Precambrian Res. 9: 93-104.

(17) Roscoe, S.M. 1973. The Huronian Supergroup, a paleoaphebian succession showing evidence of atmospheric evolution. In Huronian Stratigraphy and Sedimentation, ed. G.M. Young, paper 12, pp. 31-47. Geological Society of Canada.

(18) Schidlowski, M. 1976. Archaean atmosphere and evolution of the terrestrial oxygen budget. In The Early History of the Earth, ed. B.F. Windley, pp. 525-535. New York: Wiley and Sons.

(19) Schidlowski, M.; Eichmann, R.; and Junge, C.E. 1975. Precambrian sedimentary carbonates: carbon and oxygen isotope geochemistry and implication for the terrestrial oxygen budget. Precambrian Res. 2: 1-69.

(20) Simpson, P.R., and Bowles, J.F.W. 1977. Uranium mineralization of the Witwatersrand and Dominion Reef Systems. Phil. Trans. R. Soc. Lond. A286: 527-548.

(21) Towe, K.M. 1978. Early Precambrian oxygen: a case against photosynthesis. Nature 274: 657-666.

(22) Veizer, J. 1978. Secular variations in the composition of the sedimentary carbonate rocks - II. Fe, Mn, Ca, Mg, Si and minor constituents. Precambrian Res. 6: 381-413.

(23) Veizer, J. 1979. Chemistry of the early oceans: implications for crustal development. RSES, Australian National University Yearbook 1979: 170-172.

(24) Veizer, J.; Holser, W.T.; and Wilgus, C.K. 1980. Correlation of $^{13}C/^{12}C$ and $^{34}S/^{32}S$ secular variations. Geochim. Cosmochim. Acta 44: 579-587.

(25) Walker, J.C.G. 1977. Evolution of the Atmosphere. New York: MacMillan.

(26) Walter, M.R.; Buick, R.; and Dunlop, J.S.R. 1980. Stromatolites 3.4 - 3.5 billion years old from the North Pole area, Pilbara Block, Western Australia. Nature 441: 443-445.

Participant List

ARNETH, J.-D.
Max-Planck-Institut für Chemie
6500 Mainz 1, F.R. Germany

Field of research: ^{13}C *and* ^{18}O *isotopic measurements of precambrian rocks*

AWRAMIK, S.M.
Dept. of Geological Sciences
University of California
Santa Barbara, CA 93106, USA

Field of research: Precambrian paleobiology

BROCK, T.D.
Dept. of Bacteriology
University of Wisconsin
Madison, WI 53706, USA

Field of research: Microbial biogeochemistry and microbial ecology

BUTTON, A.
Dept. of Geology and
Geological Engineering
S. Dakota School of Mines
and Technology
Rapid City, SD 57701, USA

Field of research: Precambrian stratigraphy, sedimentation, and mineral deposits

CLOUD, P.
Dept. of Geology
University of California
Santa Barbara, CA 93106, USA

Field of research: Interactions and interfaces in biology and geology

COHEN, Y.
H. Steinitz Marine Biology Lab.
The Hebrew University, P.O. Box 469
Eilat, Israel

Field of research: Anoxygenic photosynthesis in cyanobacteria: microbial sulfur cycle in marine systems

COOK, P.J.
Research School of Earth Sciences
Australian National University
Canberra, A.C.T. 2601, Australia

Field of research: Phosphorites

CURTIS, C.D.
Dept. of Geology
University of Sheffield
Sheffield S3 7HF, England

Field of research: Sediment geochemistry

EGLINTON, G.
Organic Geochemistry Unit
University of Bristol
Bristol BS8 1TS, England

Field of research: Organic chemistry

EUGSTER, H.P.
Dept. of Earth & Planetary Sciences
Johns Hopkins University
Baltimore, MD 21218, USA

Field of research: Geochemistry

FOLINSBEE, R.E.
Dept. of Geology
University of Alberta
Edmonton, Alberta T6G 2E3, Canada

Field of research: Mineral resource economics

von GEHLEN, K.
Institut f. Geochemie, Petrologie
und Lagerstättenkunde
Universität Frankfurt
6000 Frankfurt, F.R. Germany

Field of research: Genesis of ore deposits, esp. Precambrian, using geochemistry (general and sulfur-isotope geochemistry)

GOLUBIC, S.
Biological Science Center
Boston University
Boston, MA 02215, USA

Field of research: Recent and fossil microbial endoliths, recent and fossil stromatolites, microbial ecology, microbial fossils

GOODWIN, A.M.
Dept. of Geology
University of Toronto
Toronto, Ontario, Canada

Field of research: Precambrian iron formation, archean crust

HAACK, U.
Geochemisches Institut
der Universität Göttingen
3400 Göttingen, F.R. Germany

Field of research: Geochemistry

HAHN, J.H.
Max-Planck-Institut für Chemie
Abt. Luftchemie
6500 Mainz 1, F.R. Germany

Field of research: Organic geochemical investigation of archean rocks, hydrocarbons in Isua and Pongola sediments

HALLBERG, R.O.
Dept. of Geology
University of Stockholm, Box 6801
113 86 Stockholm, Sweden

Field of research: Biogeochemistry of sediments, sedimentary sulfide formation, global sulfur budget, cycling of heavy metals at the sediment/water interface

HAYES, J.M.
Dept. of Chemistry
University of Indiana
Bloomington, IN 47405, USA

Field of research: Isotopic biogeochemistry

HOEFS, J.W.
Geochemisches Institut
der Universität Göttingen
3400 Göttingen, F.R. Germany

Field of research: Stable isotope geochemistry

HOLLAND, H.D.
Dept. of Geological Sciences
Harvard University
Cambridge, MA 02138, USA

Field of research: Evolution of the atmosphere and oceans

HOLLERBACH, A.
Lehrstuhl für Erdöl und Kohle
RWTH
5100 Aachen, F.R. Germany

Field of research: Organic geochemistry of sediments, analysis of crude oils and coals

JAMES, H.L.
1617 Washington St.
Port Townsend, WA 98368, USA

Field of research: Precambrian history, precambrian sedimentary iron deposits

JENKYNS, H.C.
University of Oxford
Dept. of Geology
Oxford OX1 3PR, England

Field of research: Pelagic sediments and authigenic mineral deposits

JUNGE, C.E.
Max-Planck-Institut für Chemie
Abt. Luftchemie
6500 Mainz 1, F.R. Germany

Field of research: The development of the earth's atmosphere

KAPLAN, I.R.
Dept. of Earth and Space Sciences
University of California
Los Angeles, CA 90024, USA

Field of research: Geochemistry

KRUMBEIN, W.E.
Universität Oldenburg
Fachbereich IV, Postfach 2503
2900 Oldenburg, F.R. Germany

Field of research: Especially laminated microbial ecosystems, geomicrobiology, cyanobacteria

LANGRIDGE, J.
Division of Plant Industry
CSIRO
P.O. Box 1600
Canberra, A.C.T. 2601, Australia

Field of research: Microbial genetics and genetic engineering

LERMAN, A.
Dept. of Geological Sciences
Northwestern University
Evanston, IL 60201, USA

Field of research: Geochemistry-sedimentology

MARGULIS, L.
Dept. of Biology
Boston University
Boston, MA 02215, USA

Field of research: Microbial ecology and evolution

McKIRDY, D.M.
Conoco Inc.
Exploration Research Division
P.O. Box 1267
Ponca City, OK 74601, USA

Field of research: Organic geochemistry of petroleum source rocks

MILLER, S.L.
Dept. of Chemistry, B-017
University of California, San Diego
La Jolla, CA 92093, USA

Field of research: Origins of life

NEALSON, K.H.
Scripps Institution of Oceanography
La Jolla, CA 92093, USA

Field of research: Marine microbiology, biogeochemistry

NIELSEN, H.
Geochemisches Institut/Isotopenlabor
der Universität Göttingen
3400 Göttingen, F.R. Germany

Field of research: Stable isotope geochemistry, especially sulfur

NISSENBAUM, A.
Academic Secretary Office
Weizmann Institute of Science
Rehovot, Israel

Field of research: Biogeochemistry

NRIAGU, J.O.
National Water Research Institute
Canada Center for Inland Waters
Burlington, Ontario L7R 4A6, Canada

Field of research: Low temperature geochemistry, environmental geochemistry

OEHLER, J.H.
Conoco Inc.
Exploration Research Division
P.O. Box 1267
Ponca City, OK 74601, USA

Field of research: Organic geochemistry in relation to petroleum exploration

PFLUG, H.-D.
Geologisch-Paläontologisches Institut
der Universität Giessen
6300 Giessen, F.R. Germany

Field of research: Precambrian microfossils

SANGSTER, D.F.
Geological Survey of Canada
Ottawa, Ontario K1A 0E8, Canada

Field of research: The nature, distribution, and genesis of stratiform lead-zinc deposits including their depositional environment

SCHIDLOWSKI, M.
Max-Planck-Institut für Chemie
6500 Mainz 1, F.R. Germany

Field of research: Geochemistry/ biogeochemistry

SCHOPF, J.W.
Dept. of Earth & Space Sciences
University of California
Los Angeles, CA 90024, USA

Field of research: Precambrian paleobiology

TRENDALL, A.F.
Mineral House
66 Adelaide Terrace
Perth, W.A. 6000, Australia

Field of research: Geology of the Hamersley Basin with particular reference to its iron formations and their significance for global precambrian evolution of the earth

TRUDINGER, P.A.
Baas-Becking Geobiological Lab.
Bureau of Mineral Resources
P.O. Box 378
Canberra City, A.C.T. 2601,
Australia

Field of research: Microbial biogeochemistry

TRÜPER, H.G.
Institut f. Mikrobiologie
Rhein. Friedrich-Wilhelm Universität
5300 Bonn 1, F.R. Germany

Field of research: Microbiology, microbial sulfur metabolism, physiology and ecology of phototrophic sulfur bacteria

VEIZER, J.
Dept. of Geology
University of Ottawa
Ottawa, Ontario K1N 6N5, Canada

Field of research: Precambrian sedimentation and low-temperature geochemistry

WALTER, M.R.
Baas-Becking Geobiological Lab.
Bureau of Mineral Resources
P.O. Box 378
Canberra City, A.C.T. 2601,
Australia

Field of research: Precambrian paleobiology and stratigraphy

WILLIAMS, N.
Carpentaria Exploration Co.
P.O. Box 1042
Brisbane, Q 4001, Australia

Field of research: The genesis of precambrian sedimentary stratiform copper, lead, and zinc sulfide deposits

Subject Index

Author Index

MIX
Papier aus verantwortungsvollen Quellen
Paper from responsible sources
FSC® C105338

If you have any concerns about our products,
you can contact us on
ProductSafety@springernature.com

In case Publisher is established outside the EU,
the EU authorized representative is:
Springer Nature Customer Service Center GmbH
Europaplatz 3, 69115 Heidelberg, Germany

Printed by Libri Plureos GmbH
in Hamburg, Germany